Biogas Plants in Europe

Solar Energy R&D in the European Community

Series E:

Energy from Biomass

Volume 6

Solar Energy R&D
in the European Community

Series E Volume 6
Energy from Biomass

Biogas Plants in Europe

A Practical Handbook

by

M. DEMUYNCK and E. J. NYNS
Unit of Bioengineering,
University of Louvain, Belgium

and

W. PALZ
Commission of the European Communities

D. Reidel Publishing Company
A MEMBER OF THE KLUWER ACADEMIC PUBLISHERS GROUP
Dordrecht / Boston / Lancaster

for the Commission of the European Communities

Library of Congress Cataloging in Publication Data

CIP

Main entry under title:

Demuynck, M.

Biogas plants in Europe.

(Solar energy R&D in the European community. Series E, Energy from biogas: v. 6)

Bibliography: p.

Includes index.

1. Biogas–Europe. I. Nyns, E. J. II. Palz, W. (Wolfgang), 1937-
III. Title. IV. Series.

TP359.B48D46 1984 665.7'76'094 84-6972

ISBN 90–277–1780–X

This Practical Handbook was prepared for the Commission of the European Communities, by a group of experts in the framework of the Solar Energy R&D Programme, Directorate-General Science, Research and Development, Brussels.
The compilation has been carried out by M. Demuynck and E. J. Nyns, Unit of Bioengineering, University of Louvain, B - 1348 Louvain-la-Neuve, under the direction of W. PALZ, from the Commission of the European Communities.

Publication arrangements by
Commission of the European Communities
Directorate-General Information Market and Innovation, Luxembourg

EUR 9096

Published by D. Reidel Publishing Company
P.O. Box 17, 3300 AA Dordrecht, Holland

Sold and distributed in the U.S.A. and Canada
by Kluwer Academic Publishers,
190 Old Derby Street, Hingham, MA 02043, U.S.A.

In all other countries, sold and distributed
by Kluwer Academic Publishers Group,
P.O. Box 322, 3300 AH Dordrecht, Holland

Printed in Great Britain

CONTRIBUTORS

This survey was commissioned for the Directorate-General for Science, Research and Development of the Commission of the European Communities, within the framework of its Solar Energy R & D Programme.

The survey was coordinated by Myriam Demuynck and Professor Edmond-Jacques Nyns (i) with Dr. Gatze Lettinga (ii) as adviser.

The survey was carried out by a Working Group of National Experts, one for each Member State of the European Community. The Working Group was composed as follows :

- Dr. Morten Riemer, Carl Bro International A/S, Granskoven 8, 2 600 Glostrup, Denmark;
- M. Dominique Verrier, Institut National de la Recherche Agronomique, Centre de Lille, Station de Technologie Alimentaire, Certia, 369, rue Jules Guesde, B.P. 39, 59 651 Villeneuve d'Ascq, France;
- M. Rolf Kloos, Bundesforschungsanstalt für Landwirtschaft, Institut für Technologie, Bundesallee, 50, 3 300 Braunschweig, Federal Republic of Germany;
- Dr. Dimitris Georgacakis, Agricultural College of Athens, Laboratory of Agricultural Structures, Iera Odos, 75, Botanikos, Athens (301) Greece;
- Dr. Emer Colleran, Department of Microbiology, University College of Galway, Galway, Ireland;
- Prof. Francesco Alfani, Istituto Principi Ingegneria Chimica, Università di Napoli, Piazzale Tecchio, 80 125 Napoli, Italy;
- Dr. A.F.M. Van Velsen, Agricultural University, De Dreijen, 12, 6703 BC Wageningen, The Netherlands;
- M. Peter-John Meynell, British Anaerobic and Biomass Association Ltd., The White House, Little Bedwyn, Marlborough, Wiltshire SN8 3JP, United Kingdom;
- Dr. Jim Coombs, Winchcombe Cottage, Bucklebury, Reading RG7 6SL, United Kingdom.

(i) Unit of Bioengineering, University of Louvain, B-1348, Louvain-la-Neuve, Belgium.

(ii) Agricultural University, 6703 BC Wageningen, The Netherlands.

It was agreed to include the Swiss experience of biomethanation in the survey. This has been made possible through the courtesy and assistance of Dr. Art Wellinger, Eidgenössische Forschungsanstalt für Betriebswirtschaft und Landtechnik, CH-8 355 Tänikon, Switzerland.

A comparative survey of the status of biomethanation in the United States was made possible through a round trip in the U.S.A. in June 1982.

The authors wish to thank :
Mr L. Crossby for editing the text in English,
Mr A. Rozzi, AC Biotechnics, AgipGiza, AIDR, CSM, RPA and the experts for the photos,
Ms C. Buydens and Mr G. Schepens for the analysis of the economic data,
Mr A. Buendia for the spanish data,
Ms E. Butaye for the word processing,
and all those, too numerous to cite individually who have made this survey possible.

All original formats and national papers have been gathered in a compendium of 1054 pages and can be ordered at cost price of the photocopy at the Unit of Bioengineering, University of Louvain, B-1348 Louvain-la-Neuve, Belgium, under No. EUR 9096.

PREFACE

The production of biogas from wet or liquid organic residues either on the farm, in the agro- and food industry or from municipal waste (landfills) has only gained real interest since the end of the seventies, after the second energy crisis. Since then, a considerable number of biogas plants of various sizes have been installed all over Europe, most of them on private initiative. At first, the aim was for energy production but de-pollution soon became an important additional motivation for building such plants.

The present book is conceived to give guidance on the state-of-the-art and possible future developments. Conclusions are drawn from more than 500 plants which were analysed for this purpose. A selection of them are described in detail as possible examples.

Reference lists of operational plants and manufacturers as of 1983 are included as well as information on governmental supports for financing.

Apart from this practical information, the book also covers problem areas and makes suggestions for future R+D.

The work for this book was initiated and directed by the Commission's staff in Brussels. Its preparation was in the hands of a European group of experts and a coordinator team in Louvain-la-Neuve, Belgium.

In addition to this book, the experts have prepared special reports for Ireland, United Kingdom, Denmark, Germany, the Netherlands, Belgium, France, Italy and Greece which are distributed by the coordinator in Louvain-la-Neuve.

W. PALZ
Head of the R+D Programme for
Solar Energy, Biomass and Wind Energy

CONTENTS

EXECUTIVE SUMMARY OF THE SURVEY

Biogas plants in the European Community and related questions

Biogas is a gaseous fuel, rich in methane produced through a biological route in a methane digester from a wide variety of substrate biomasses, including organic wastes and residues. In other words, biomethanation is a process which takes place in a biogas plant. A total of 546 methane digesters, built in the European Community and Switzerland during the last five years, were identified in this study. The principal questions arising in connexion with biogas plants may be summarized as follows :

- What is a biogas plant ? On which biological process is it based ? What technology(ies) does it involve ?

- What are the potentialities of biogas plants in the European Community Member States and in Switzerland ? Will they contribute significantly to the energy demands of the European Community ?

- Is it economical to build a methane digester ? Do the public authorities subsidize construction and/or follow-up ? What is the pay-back period for the investment ? Are there fringe benefits ? What is the legal context in which this takes place ?

- Is the present process of biomethanation productive, reliable ? What are the engineering problems of a biogas plant ? What are the operating problems ?

- How can the bottlenecks be overcome ? What incentives are given by the public authorities such as the Commission of the European Communities, to help overcome the bottlenecks ? Should this objective be realized, what are the prospects for the implementation of biogas plants in the European Community ?

The assessment of "Biogas Plants in Europe" sponsored by the Commission of the European Communities

The foregoing questions prompted the Commission of the European Communities to originate the present assessment of "Biogas Plants in Europe" in the European Community Member States and in Switzerland in order to obtain, first, an accurate knowledge of the present status, together with the possibilities and limits of the technology of biomethanation and, secondly, a clear view of the problems of biomethanation which must be solved if this technology is to be implemented on a wider scale.

A survey was carried out from mid-1981 to the end of 1982 by a team of national experts. Each of them identified, visited and scrutinized almost every existing methane digester in his home country, following a common format. This book presents a synthesis of the survey which tends to answer the questions posed above. The information has been updated to March 1983 on the basis of further data on new biogas plants provided by the national experts.

Quantitative data on existing biogas plants

The 546 biogas plants identified at the beginning of 1983 in the European Community Member States and in Switzerland represent about 95 000 m3 of methane digester working volume on farms and 174 000 m3 of methane digester working volume in industry, making a total capacity of 269 000 m3.

On the basis of normal running conditions of methane digestion and, hence, a theoretical mean methane productivity of 1 m3 biogas per m3 digester working volume per day, a daily output of 269 000 m3 of biogas may be expected. If we assume that, on average, 30 % of the biogas produced is needed to run the biogas plant itself and that the remaining 70 % are effectively used, this represents already in 1983 a yearly output of 33 000 tonnes oil equivalent (t o.e.), worth 7 million European Currency Units (ECU)(1).

From the evolution of the number of methane digesters as a function of time, it appears that the learning curve for biogas plants on farms started in 1979-1980, followed by the learning curve for biogas plants in industry which started in 1980-1981. (For example, in Greece, where research in the field of biomethanation started only in 1981-1982, 4 full-scale methane digesters were operational by 1983). Municipal wastes are often disposed of as landfills which can be transformed easily into "natural" methane digesters. The learning curve for the latter biogas plant system in the European Community started later, in 1981-1982, but interest in the abstraction of biogas from landfills is evolving rapidly. Whereas there were only 2 sites in operation in 1980, a total of 36 sites can be identified in 1983 in 5 of the States covered by the survey.

Potentiality of biomethanation in the European Community

The potential contribution of biomethanation to the energy requirements of the European Community has been evaluated. Biomethanation of animal wastes can provide yearly 14.1 Mt o.e., crop residues 8.5 Mt o.e. (Palz and Chartier, 1980), and municipal wastes 8.9 Mt o.e. (Hall, 1983). The total potential production of 31.5 Mt o.e. represents only 3 % of the present estimate of European energy consumption in 1985, but it has a market value of 7 000 million ECU. This means an important but local potential. When it becomes customary to grow energy crops as catch crops, algae or aquatic plants these values may increase threefold (Hall, 1983).

A wide variety of wastes can be biomethanized

The number of wastes which are suitable for biomethanation is considerable, and a wide variety of wastes are already being used. Thirty-seven

---

(1) One ECU is roughly equivalent to one U.S. dollar.

different agricultural wastes have been identified in this study. They include all types of manure, with or without bedding, and some crop residues which are biomethanized either alone or in various combinations. Energy crops for biomethanation are not yet popular either within the European Community or in Switzerland. Only one pilot biogas plant, using algae as biomass substrate, has been identified in the south of Italy. Twenty-three different industrial wastes are already used as biomass substrates for biomethanation. Among these, two are solid wastes, originating from sugar refineries and slaughter-houses respectively. The 21 others are wastewaters, most of which originate in the agro-food industries.

The numerous methane digestion systems

Many methane digestion systems have been designed and adapted to the different types of wastes to be biomethanized. For slurries such as pig or cattle manure, the one-step, once-through, completely mixed system is in general use. For solid wastes, such as manure with bedding, the old, well known batch system is still employed. For agro-industrial wastewaters, on the contrary, more advanced methane digestion systems are used in which the bacteria are either trapped or immobilized. Typical examples are the anaerobic filter and the upflow system without carrier.

Most methane digesters identified in the survey operate under mesophilic conditions, that is around 35 °C. There is a tendency in Germany and Switzerland to operate the methane digesters at ambient temperature and this can prove economical. Thermophilic conditions, that is temperatures around 55 °C, are rarely encountered in practice.

One-third of the biogas plants in Europe are "do it yourself" plants, that is they were constructed by the owners with or without external assistance. Two-thirds of the plants are "turn-key" plants, that is they were constructed by specialized companies or local tradesmen, but this does not exclude active collaboration by the owner, a case often observed in Denmark.

Economics of biogas plants

Only limited information is available at present concerning the economics of the biogas plants existing at the end of 1982 in the countries covered by the survey. Not many methane digesters have been in routine operation for more than one year and, if so, their economics have not always been properly evaluated. Nevertheless, valuable data were obtained for 32 biogas plants on farms.

For this economic analysis, the Financial Model Processor of Schepens, (Namur, Belgium) has been used. This model is based essentially on the concept of cash-flows. Clearly, a biogas plant can already be economical. Indeed, of the 32 biogas plants scrutinized, 6 have Simple Pay-Back Periods for the investment ranging between 3 and 6 years. Seven plants have Simple Pay-Back Periods for the investment between 7 and 9 years, and 19 plants have Simple Pay-Back Periods equal to or greater than 10 years.

Why is it then that most of the methane digesters on stream in 1983 appear to be uneconomical ? Two major causes were found, as discussed below. In some particular cases, there were also other reasons such as poor gas utilization, high maintenance costs and an unduly large requirement for the heating of the digester. It will be seen that none of these reasons is intrinsic either to the process or to the technology, and all can be overcome.

The first main cause of the non-profitability of a biogas plant is poor operation of the methane digester. An average daily productivity of 1 m3 of biogas per m3 of methane digester is the lower limit for economical operation. That productivities of this order and higher can be obtained has been widely demonstrated, even for full-scale methane digesters. It may be recalled here that the average productivities reported for methane digesters have increased threefold between the first European Community Congress on Biomass in Brighton in 1980 and the second European Community Congress on Biomass in Berlin two years later (Chartier, 1983). However, few biogas plants meet this average daily productivity of 1 m3 per m3 methane digester in practice and the reasons for this are analysed below. Moreover, all the biogas produced should be used effectively, but this is not often the case for a number of reasons which will be analysed below. In passing, it may be noted that problems, both legal and financial, still arise when effective use of the produced biogas involves sale to third parties.

The second main cause of the non-profitability of a biogas plant is high investment cost. To be economic, the investment cost for the overall biogas plant must not exceed 400 to 450 ECU per m^3 digester working volume. Although the building costs of some existing biogas plants have been found too high, and therefore appear as a reason for economic failure, these cost limits for investment can generally be met in the countries surveyed in this study.

Motivations for building a biogas plant

An obvious but real motivation for building a biogas plant is the local availability of large quantities of wastes or residues, but other factors may positively influence the decision to build a biogas plant. These factors do not necessarily involve energy production, one of them being the legislation concerning pollution control. Italy is a typical case. Large piggeries situated in the northern part of that country may no longer discharge their wastewaters or waste sludges unless they pay for depollution. This explains the wide distribution of biogas plants in the north of Italy. However, the added value of depollution is as yet still difficult to evaluate properly.

Less obvious but just as real is the local influence of research centres active in the field. A good example is the Landtechnik Weihenstephan Institute, located near Munich, which promoted not only the first but a number of other biogas plants in the Federal Republic of Germany.

Bottlenecks in the implementation of biomethanation in Europe and Switzerland

Bottlenecks in the implementation of biogas plants in the countries surveyed by this assessment and which presently deter both farmers and industrialists from investing in a biogas plant are of two orders. The first is the general context in which biomethanation is developing. Legislation for controlling pollution is at present limited. Where such legislation does exist, e.g. in The Netherlands, the aerobic biological treatment of wastewaters remains an actively competing alternative to anaerobic biomethanation. The recent highly productive biomethanation technologies have as yet few working full-scale examples or even remain to be demonstrated at the full-scale level. In the same general context, in many countries there is only scarce financial backing for building a biogas plant. Especially farmers are deterred because the agricultural sector is suffering great financial difficulties.

The second order of bottlenecks is related more particularly to the biomethanation technology itself. For a number of feedstocks potentially available for biomethanation, there still does not exist a reliable methane digestion system. This is especially true for bedding manure, although some of the early biogas plants already dealt with this biomass substrate. So far as the engineering, the operation and the monitoring of biogas plants are concerned, a number of problems were highlighted by the survey. Many methane digesters still exhibit a lack of reliability, especially during the first year of operation. For example, from the detailed information obtained for 96 biogas plants, it appeared that each plant has suffered an average of 2 major problems (temporary shutdown of the plant or non-reversible souring of the methane digestion mixed liquor) and 4 minor problems (with no major disruption of operation of the plant).

The most important problems in the engineering and maintenance of biogas plants are the following : gas leakage of the methane digesters, failure of the feedstock pumping device, and inefficiency of the mixing systems. Concerning gas utilization, particular problems are arising with electricity generation systems. Simple and robust generating sets seem to be a necessity but are still lacking. As regards the operation of biogas plants problems are occurring with settlement and water entry during feedstock storage. The loading operation presents problems with blockage of pipes and heating. The digester operation itself presents problems with settlement and/or scum formation. Finally, concerning the monitoring of biogas plants, gas metering appears to be the most important problem.

Scenario for implementation of biogas plants

If we assume that the bottlenecks identified above will be overcome, that legal policy for the management of the environment will be reinforced, and the tensions in the energy market at the political and economic levels will persist, the scenario for implementation of biomethanation could be as follows. For biomethanation on farms, it can be expected that, within the next 10 years, methane digesters for the treatment of manure will be constructed in series, so that they will become economical irrespective of the number of livestock units and hence suitable for a wide size range of farms. In addition, simplified and robust methane digesters will become available as a "do it yourself" technology and progressively extend the market to farms of all sizes. Later progress brought about by R & D projects will open the market to solid agricultural wastes also, e.g. bedding manure, as feedstock for biomethanation.

In industry, biomethanation and particularly the improved, second generation technologies will within the next ten years, gain general acceptance for agro-industrial wastewater treatment. Progressively, it can be expected that biomethanation will be applied to a larger range of industrial feedstocks, including solid industrial wastes and/or wastes of non-agro-food origin. In these cases, the methane digester will usually be integrated as a unit operation in a global treatment technology.

As regards the developing countries, family-size methane digesters with appropriate designs adapted to local conditions will certainly contribute to the local energy supply and autonomy, especially in rural areas, following the example of China. In this particular case, the possible use of the solid portion of the residue resulting from biomethanation as compost, soil conditioner or fertilizer may even become the major incentive. Moreover, as in industrialized countries, the agro-industries in developing countries will incorporate biomethanation so as to protect the environment and save energy.

Whereas the above developments are expected in the next ten years in existing farms or plants, the new farms and agro-industries which will be better integrated into their environment will later prove to be partly or even totally energy self-sufficient.

The use of urban refuse for biomethanation is expected to develop rapidly. An increasing number of landfills will be exploited for biogas extraction. Later it can be expected that urban waste dumps will become the starting point of biogas refineries. Other industrial and/or agricultural wastes may be collected and biomethanized together with domestic refuse so that these biogas refineries will become a potentially important complementary source of town gas.

How can the bottlenecks of biomethanation be overcome ?

The following recommendations are put forward to overcome the bottlenecks and to implement the biotechnology of biomethanation. First and foremost, a European network of selected biogas plants should be set up. The monitoring of these plants, together with the collection of data on engineering, biological and economic aspects will permit a substantial popularization of the biomethanation process so that a confident market for biogas plants could be promoted. Within this European network, the use of a common language needs to be promoted and especially the standardization of methods, parameters, symbols and units.

Secondly, because there is a need for a better understanding of the basic processes on which biomethanation relies, basic research should be promoted so as to improve yields and rates in existing processes, to include new biomass substrates, and to obtain a larger variety of energy-yielding molecules.

Thirdly, in order to improve the engineering aspects of biomethanation, R & D projects should be aimed at the comparative evaluation of reliability with time of a wide range of equipment and the development of cheap, reliable methane digesters. R & D projects should also aim at improving the basic technology for second-generation methane digesters, including mathematical modelling.

The fourth recommendation is that the integration of methane digestion systems in the overall planning of the farm, including the feasibility of collective biogas plants, should be improved with a view to achieving self-sufficiency in energy.

Finally, a closer look at the upgrading of side-products of the biomethanation process will give an added value to these products and could provide a valuable incentive. Finally, the development of safety legislation and standardization should also promote the potential of biomethanation.

AN INTRODUCTION TO BIOMETHANATION AND BIOGAS PLANTS

0.1. The need for new fuels

The dependence of the European economies on the continuous availability of fossil fuels was brought out sharply during the energy crisis of the early seventies. Europe realized then how urgent it was to seek new sources of energy and a greater degree of autonomy in the energy field. Along with the drive for economies in energy consumption, there emerged at that time a growing interest in the use of infinitely renewable sources of energy. Harvesting the sun through biomass, for example, began to look more and more attractive.

Two facts reported to the 2nd European Community Conference "Energy from Biomass", held in Berlin in September 1982 should be mentioned here. The first is that the resolution of the European Parliament on the use of biomass as a source of energy was unanimously adopted in September 1982. The second is that biofuels are now being considered within the strategy of the public authorities in charge of agricultural policy within the European Community (Chartier, 1983).

Simultaneously with the trend towards a new energy mix, there has been a great effort towards environmental protection almost everywhere in Europe. All this explains the interest since 1975 for biomethanation or anaerobic digestion, a biotechnology by which an organic waste, to be disposed of in any case, can be recycled at least partly as energy, i.e. biogas. The process is a microbial anaerobic fermentation which takes place in a closed vessel or methane digester. The latter in its surroundings will be referred to as a biogas plant.

0.2. The history of biogas

McCarty (1982) recalled that the formation of methane from anaerobic decomposition of organic deposits was known as early as the eighteenth century, following its discovery by the Italian physicist Alessandro Volta in 1776. Nearly a century later, in 1881, anaerobic treatment was reported to be a useful method for reducing the mass and the putrescible nature of suspended organic material removed from municipal wastewater. The first "air-tight chamber" was developed in France by Louis Mouras for that purpose. The history of biogas is described at length in a book by Van Brakel (1980) entitled "The Ignis Fatuus of Biogas".

Interest in biomethanation existed before and after the Second World War, when indigenous energy sources were scarce. In France, the pioneering work of Ducellier and Isman (1941) and of Sauze (1947) led to the proliferation of "do-it-yourself" small-scale biogas plants. Fig. 0.1 illustrates such a rural batch digester. In Germany, several improved or new designs for biogas plants were developed : the "Darmstadt" system (Reinhold and

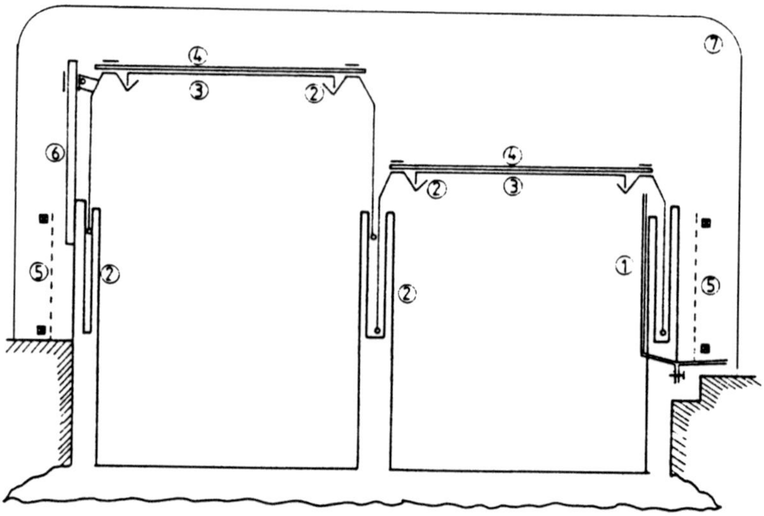

Fig. 0.1. Rural batch digester

1 : Gas outlet; 2 : Water seal; 3 : Cover; 4 : Fixation bar; 5 : Mixed straw and manure insulation; 6 : Gasometer guide; 7 : Glass housing for insulation (from Ducellier and Sauze, 1975)

Noack), the double batch system, and the "Berlin" system (Ikonomoff and Gärtner). Strell, Liebman and Göts in Munich, as well as Müller in Hohenheim and Poetsch in Hanover tested biogas plants for the methane digestion of solid manure (Baader and Kloss, 1982).

With the coming of cheaper fuels during the fifties and sixties, activities in the field of biomethanation decreased rapidly. Most of the biogas plants built by that time were progressively shut down. Only primary sludges and excess activated sludges, originating in wastewater treatment plants, were still subjected to biomethanation. In fact, this process has been in regular use for the stabilization of these sludges on a very large scale for many decades and is still popular today. In the Federal Republic of Germany for instance, more than 60 % of the excess sludges from wastewater treatment plants are stabilized in more than 900 biogas plants in operation (Baader and Kloss, 1982), representing in 1979 a capital investment around one thousand million DM.

After the oil crisis of 1973, a revival of interest in biomethanation evolved rapidly in the different countries of the European Community. The development of "green routes", initiated by ecological pressure groups, prompted many people to look once again at methane digesters. A typical case is the once-through, completely mixed continuous methane digester of 75 m^3 working volume, built on the farm of M. Steiner in 1976 at Montherod, Switzerland, and operating on cattle manure. Research and development in

Allerhop (F.R. Germany) 1950. General view of the biogas plant. Picture from Def Mitteilungen, Heft 9, 1951. Published in Verden.

Allerhop (F.R. Germany) 1950. Biogas is compressed in high pressure bottles for use as tractor fuel. Picture from Def Mitteilungen, Heft 9, 1951. Published in Verden.

biomethanation as well as the building of full-scale biogas plants followed rapidly.

The process of biomethanation appeared relevant not only for the classical agricultural wastes but also for the treatment of raw wastewaters. Abstraction of biogas from landfills, i.e. tips where domestic refuse is disposed of, started somewhat later in Europe; this will be discussed in another chapter.

0.3. What is a biogas plant ?

The heart of a biogas plant is of course the methane digester itself. However (Fig. 0.2), upstream, the biomass resource has to be stored and/or pretreated before biomethanation. Downstream, the biogas produced and the effluent from the methane digester must be stored and/or post-treated. Furthermore, the biogas plant must be correctly integrated in its surroundings and therefore, correct links with the biomass resource upstream and with biogas and effluent uses downstream are essential.

Most of the wastes produced on a farm or by industry need to be stored before they are used or disposed of. Then why not take advantage of this necessary storage to biomethanize the waste before its further utilization ? In fact, a methane digestion tank is nothing more than a closed storage tank producing methane-rich biogas which can replace part of the energy needs of the farm or industry.

This concept explains why some Swiss farmers transformed their manure storage tanks, located underneath the stables, into methane digestion tanks. The manure falls directly by gravity into the methane digester, which needs little or no heating. Biogas is produced and used on the farm.

Where there is no possibility of transforming the storage tank directly into a methane digester, the tank can nevertheless be used, besides its storage function, for pretreating the waste prior to biomethanation. For example, a tearing pump is often placed inside the storage tank of manure, so that straw and other rough suspended solids can be shredded and mixed with the manure.

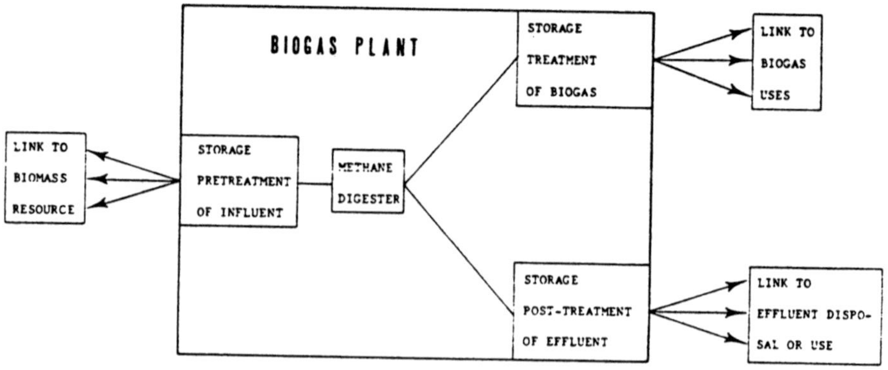

Fig. 0.2. Flowsheet of a biogas plant

For industrial wastewaters, the storage tank also serves as buffer tank to equalize wastewaters from different origins or varying with time. These buffer tanks are also used for pretreatment such as pH correction by addition of calcium carbonate or chalk.

The methane digester itself (Fig. 0.3) must be constructed as near as possible to the site of production of the waste or near the storage tank so as to minimize load losses. The tanks are generally cylindrical, whether on farm or in industry. A variant, often encountered in Germany is the egg shape, used for methane digesters of sewage sludges.

The methane digestion tank must be a closed vessel which is gas-proof, at least in its upper part. For example, concrete is currently used for the construction of such tanks. The concrete must be protected on the inside by a gas-tight coating such as epoxy paint.

As will be seen below, the biomethanation process rarely occurs at ambient temperature, consequently the methane digester must be heated to an appropriate temperature. A heating system is needed which will heat the feedstock that enters the methane digester and also maintain the desired temperature in the methane digester. Internal or external heat exchangers are often used for this purpose. Also, to optimize efficiency, it is very important to reduce heat losses from the methane digester by insulating the tank walls. All types of insulation materials, such as are used for the insulation of houses, may be employed. The thickness of the insulation will of course depend on the ambient temperatures : a greater thickness is required in Denmark than in the south of Italy. It is also extremely important to coat this material with a waterproof cladding, since any ingress of moisture drastically reduces the insulation properties.

In addition to the possible heating system, the methane digester will normally contain a mixing device. Propellers or paddles may be used for mechanical mixing. Several other more sophisticated mixing systems are encountered in practice. This will be discussed in detail in a later chapter.

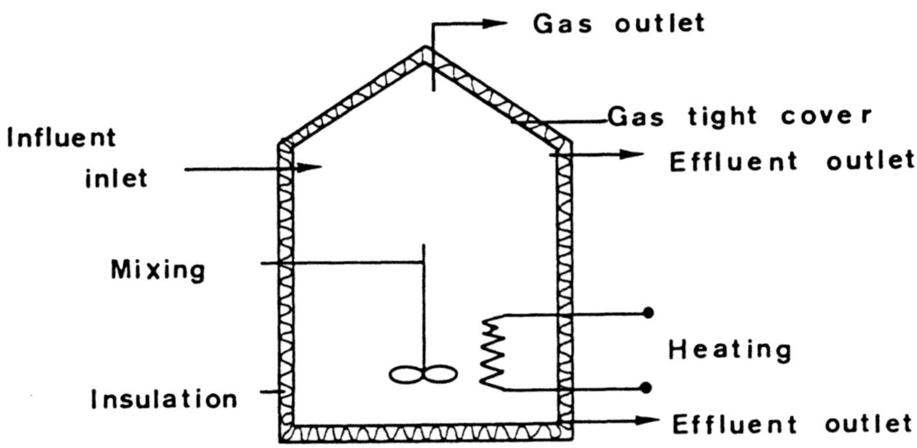

Fig. 0.3. Diagram of a methane digester

Once biomethanized, the digested waste produced daily must be stored, awaiting its use on the land or disposal. The digested sludge is "stabilized" and therefore deodorized, hence simple lagoons are sufficient for storage. When further depollution is required or when no land is available a post-treatment system is integrated in the biogas plant. A typical example is provided by digested sewage sludges which often need to be dewatered in thickeners and then subjected to mechanical pressing.

The second product of the methane digester process is of course the biogas. Because its methane content varies between 55 and 75 % (vol vol^{-1}), the biogas has a calorific value ranging between 20 and 26 MJ m^{-3} and it may thus be used to replace some of the energy needs of the farm or industry. However, in most cases, the biogas demand does not coincide with the pattern of production so that an intermediate gas storage is necessary. Storage however, makes it possible to control the gas pressure. Two types of storage may be employed. The first is the variable pressure and fixed volume type such as the high pressure bottles used for Light Petroleum Gas (L.P.G.) storage. The second is the variable volume and fixed pressure type such as gas balloons or floating gasholders.

When the biogas comes out of the methane digester it contains not only methane but also water vapour, carbon dioxide (CO_2) and small amounts of hydrogen sulphide (H_2S) and dihydrogen (H_2). The water vapour can be removed by condensation. The hydrogen sulphide is trapped easily by an iron oxide sponge. Scrubbing of carbon dioxide is unnecessary for most biogas uses. This is nevertheless feasible, although it is expensive.

Biogas can be used for heat production, electric power production or driving vehicles. The pattern of biogas utilization will of course depend primarily on the particular energy needs of the biogas production site, but it can also be influenced by special situations. Dairy farms will prefer electricity while piglet breeding units will prefer heating. The biogas produced from sugary wastewaters is generally used for pulp drying.

In some situations, exchange agreements can be made between the biogas plant and a nearby industry. The best example of this is the biogas plant of the Frauenthal cloister in Switzerland : when there is an excess of biogas production, the biogas transformed into electricity is loaned to a nearby paper mill; when there is a demand for electricity on the cloister farm, the paper mill loans it back. Another typical example is provided by the sewage treatment works at Colchester, Essex, United Kingdom, which uses its surplus biogas as fuel for road vehicles.

Three books deal at length with the biomethanation process : one in French, written by Lagrange in 1979 and two in English, one written by Stafford et al. in 1980 and another edited by Wise in 1981.

0.4. Biomethanation, a biological process

Now that we have seen how the biomethanation technology is applied in the biogas plant, it is time to take a closer look at the biological processes going on inside a methane digester. Biomethanation or more precisely methanogenesis is a microbial process which involves several steps, each of them governed by a particular bacterial community.

The example of what happens to a crop residue during methanogenesis will permit a better understanding of the biological process (Fig. 0.4). When the crop residue is kept in pile for one or two days, there occurs a liquefaction caused by hydrolytic breakdown. The rough solid organic matter is decomposed into soluble small molecules. This first step of hydrolytic liquefaction is mediated by a microbial community of fermentative bacteria which comprises both strictly and facultative anaerobic bacteria. First, due to enzymes located on the outer side of the bacterial cell wall or released by the bacteria in the outer medium (these enzymes are therefore called exo-enzymes), the solid crop residue is broken down, i.e. hydrolysed into soluble small molecules. E.g., cellulose is transformed into soluble sugars such as glucose or cellobiose. These soluble molecules, in turn, are taken up and used by the former and other fermentative bacteria for their growth which results in the release of reduced end products in the outer medium. The latter are lower molecular weight alcohols such as ethanol, lower molecular weight acids such as lactic acid or volatile fatty acids from two to five carbon atoms (acetate, propionate, butyrate...). Bicarbonate, HCO_3^-, and dihydrogen, H_2, are also produced.

All products or metabolites from the former liquefaction-fermentation step other than acetate, CH_3COO^-, bicarbonate, HCO_3^-, and dihydrogen, H_2 need further transformation before they can effectively yield methane. It is the role of a number of strictly anaerobic bacterial species belonging either to the obligate hydrogen producing acetogens (OHPA) or to the sulphate reducers, to transform or metabolize these products or metabolites into acetate, CH_3COO^-, bicarbonate, HCO_3^-, and dihydrogen, H_2.

Finally, two types of methanogenic bacteria take over. The first, called acetoclastic, dismutates acetate, CH_3COO^-, into methane, CH_4, and bicarbonate, HCO_3^-. The second, called hydrogenotrophic, reduces bicarbonate, HCO_3^-, through dihydrogen, H_2, to methane, CH_4. As a result of the former metabolism and of physico-chemical equilibria in the liquid medium, the biogas produced contains usually between one half and three quarters of methane, CH_4, and between one quarter and one half of carbon dioxide, CO_2.

The methanogenic bacteria isolated hitherto all belong to a strange class of bacteria : the archaebacteria. These bacteria are the oldest creatures still living on the Earth.

Not all wastes have to undergo all the steps depicted in Fig. 0.4. E.g. many wastewaters of the agro-food industry already contain mainly soluble molecules and consequently enter the metabolic scheme of methanogenesis at a later stage, a fact which is reflected in the design of the biogas plant.

0.5. Environmental factors in methanogenesis

As in all biological processes, the environmental conditions, i.e. the ecosystem in which the reactions take place, greatly influence the yields and rates of the biomethanation process. Two types of factors will be considered : first, the physical factors which include temperature and mixing; secondly, the chemical factors which include pH, the redox potential, and the carbon to nitrogen, C/N ratio.

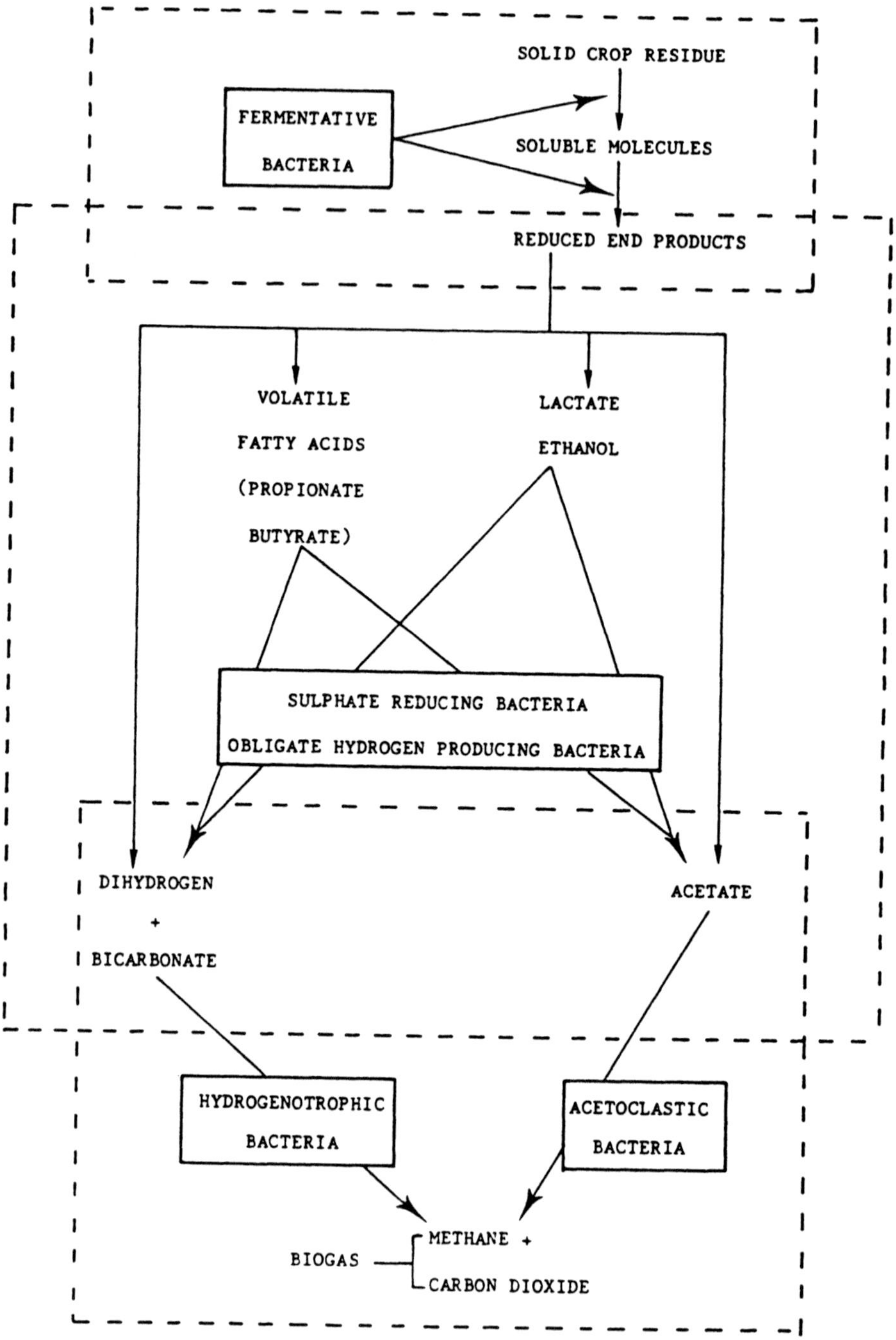

Fig. 0.4. Flowsheet of the biological process involved in biomethanation

0.5.1. Physical factors

a) Temperature

The process of methanogenesis can occur in a wide range of temperatures : from 4 °C, as in lake sediments to well above 60° to 70 °C. Zeikus has even discovered a methanogenic bacterial species acclimated in hot springs at 85 °C.

As in many if not all biological processes, the rate of activity increases with increasing temperature. Consequently it does not usually pay to run a methane digester under psychrophilic conditions, i.e. at ambient temperature and below. In fact, most European methane digesters operate in the mesophilic range of 30° to 40 °C, and only a small number of plants operate in the thermophilic range of 50° upwards.

Nevertheless, there is an intermediate range between 42° and 50 °C where no particular bacterial species is favoured. It is thus not surprising that no methane digester operating between 38° and 50 °C was found in Europe, as will be noted in the survey.

Thermophilic methanogenesis is less utilized, essentially because more energy is needed to heat up the influent load and to maintain the thermophilic temperature within the methane digester. Even within the most widely used mesophilic range, some methane digesters operate around 30 °C rather than at 35 °C. This is often the case in The Netherlands.

b) Mixing

The effects of mixing the content of the methane digester are not yet well known. Mixing is, sensu stricto, not necessary for the methanogenic biology to take place. It does not greatly modify the biogas productivities, namely the production rate and yield, or biogas composition. It certainly achieves a uniform temperature inside the methane digester and reduces or even eliminates scum formation and settlement inside the digestion tank.

0.5.2. Chemical factors

a) pH

The bacterial community responsible for methanogenesis requires a neutral pH, i.e. around 7. Below pH 6 and above pH 8, methane digestion falls off rapidly. The pH of the methane digestion liquor is a function of the "alkalinity", a complex parameter which takes into account, among other factors, the concentration in volatile fatty acids and the presence of bicarbonate, HCO_3^-.

When the concentration in volatile fatty acids rises too high, the buffering capacity of the bicarbonate anions becomes insufficient to maintain the pH level around and above 7. A concentration range of bicarbonate between 2.5 and 6 g l^{-1} usually provides a sufficient buffering capacity.

b) The redox potential

The redox potential is a complex function of soluble chemical compounds present in the digestion mixed liquor as products, metabolites or substrates. The methanogenic bacteria require a very low redox potential - of the order of - 250 mV or below - to live and produce methane. The redox potential is given by the couple methane, CH_4, and carbon dioxide, CO_2, and by the couple proton, H^+, and dihydrogen, H_2.

If, in the digestion mixed liquor, there are other buffering redox couples which lead to a higher redox potential, the activity of the methanogenic bacteria slows down. This occurs, e.g., when some oxygen or air is introduced accidentally in the methane digester.

c) Carbon to nitrogen ratio

The microbial population involved in anaerobic digestion requires sufficient nutrients to grow and multiply. In this respect, a maximum 100 : 3 carbon to nitrogen ratio (wt wt^{-1}) is necessary to maintain bacterial growth. This arises from the elemental composition of the bacterial cells which is roughly $C_5NH_7O_2$, and from the fact that 15 % of the carbon substrate is assimilated by the bacteria.

However, when there is too much nitrogen in the substrate biomass, the latter accumulates as ammonia, NH_3/NH_4^+, in the methane digestion liquor to amounts which become toxic for the methanogenic bacteria. This was encountered in a methane digester running on pig manure and located in Klarenbeek (The Netherlands). In this case, in order to avoid toxic concentrations of ammonia nitrogen above 5 000 mg l^{-1} in the digestion mixed liquor, the pig manure had to be diluted before it was fed to the methane digester.

0.6. Biomethanation systems and types of biogas plants

Table 0.1 summarizes the principles of the existing biomethanation systems.

0.6.1. Batch systems

The early biomethanation systems were all batch digesters. The waste to be treated is placed in a closed box and kept there for a long time, say from 8 weeks to several months. Table 0.1 depicts the batch and the fed-batch systems. A typical example of the batch system is digester 315 in Marchesieux (France) (see Part I, Chapter I). A typical example of the fed-batch system is digester 256 in Dreien (Switzerland) (Part I, Chapter I).

The problem with batch methane digesters is that the biogas production rate is very irregular, high at the beginning of the digestion period and very low near its end. Moreover, the batch systems require, at times, a large input of labour. Nevertheless, the batch system remains appropriate for the biomethanation of solid residues. In France, for example, there are about 20 batch methane digesters on manure with bedding.

Table 0.1. Principles of existing biomethanation systems

Biomethanation system	Principle	Diagram
Batch system	Biomethanation system characterized by discontinuous loading without elimination of the digester content. The simple batch system has a single feeding and may have a mixing system.	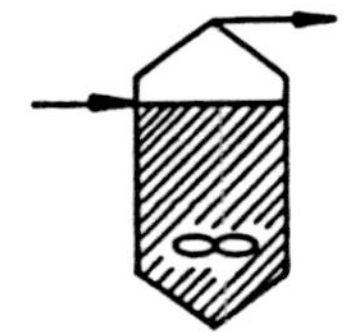
Fed-batch system	Biomethanation system characterized by discontinuous loading without elimination of the digester content. The fed-batch system has multiple feedings (a,b,c,d).	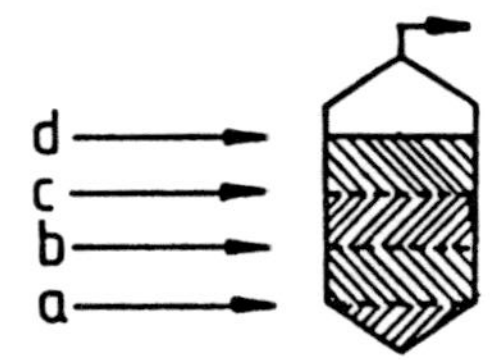
Completely mixed continuous system	One-step, once-through biomethanation system with a mixing system which homogenizes the digester content. The loading can be continuous or intermittent.	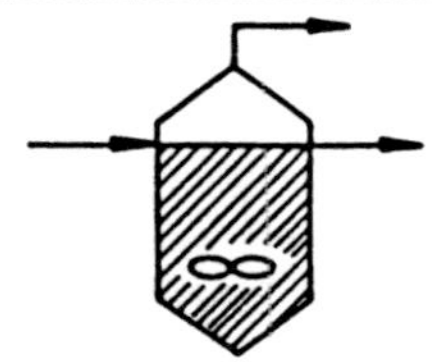
Partially mixed continuous system	Differs from the completely mixed system by incomplete homogenization of the digester content.	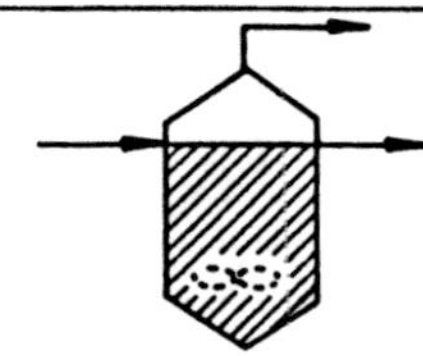

Table O.1. Principles of existing biomethanation systems (continuation)

Biomethanation system	Principle	Diagram
Plug-flow continuous system	One-step, once-through biomethanation system with horizontal flow, which may be mechanically assisted.	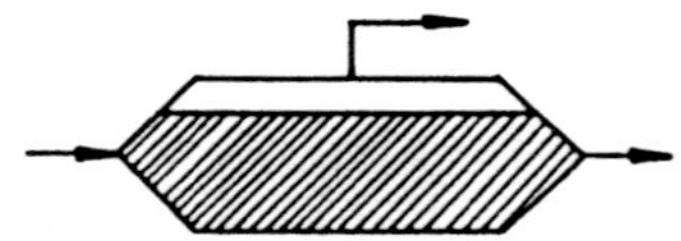
Anaerobic contact	Same as the completely mixed continuous system, but with recycle of the active biomass after settling.	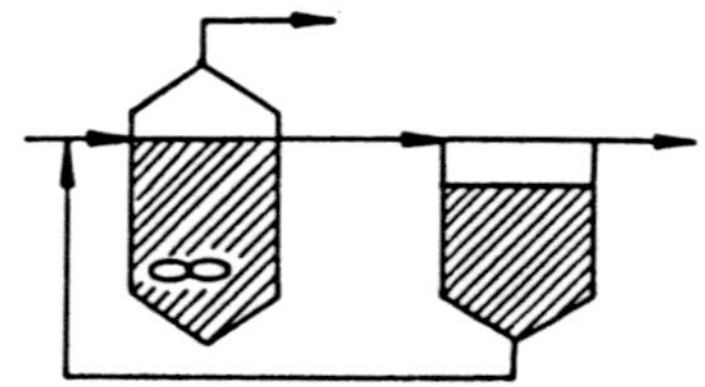
IRIS system	System developed by the Institut de Recherches de l'Industrie Sucrière. Similar to anaerobic contact but in one tank.	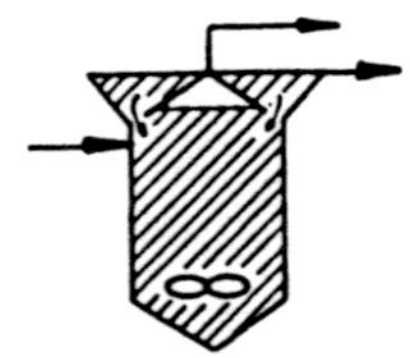
Upflow sludge blanket (upflow without carrier)	Upflow methane digester with an incorporated settler/gas separator. The settled sludge remains as a sludge blanket in the lower part of the digester.	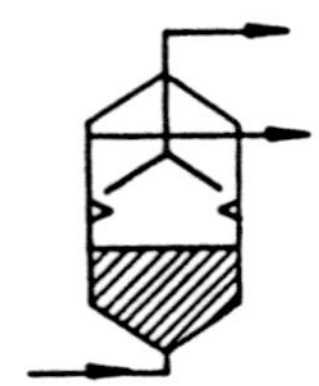

Table 0.1. Principles of existing biomethanation systems (end)

Biomethanation system	Principle	Diagram
Anaerobic filter	Up- or downflow process with trapping and immobilization of the active biomass on a fixed carrier (fixed film).	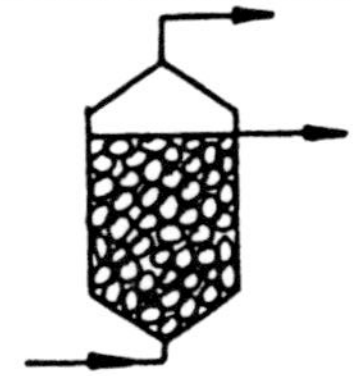
Fluidized bed	Upflow process with trapping and immobilization of the active biomass on small micro-carriers which are kept fluidized by liquid recycle.	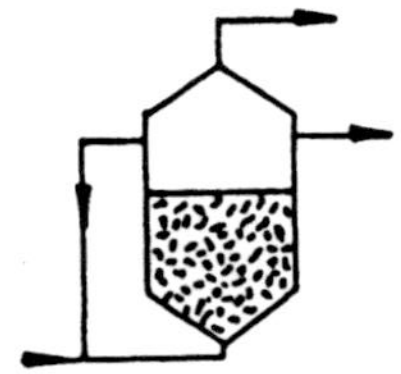
Two-step	Biomethanation in two separate reactors in series. Liquefaction fermentation (acidogenesis) occurs in the first reactor whereas methanogenesis occurs in the second reactor. Each can be designed according to the above principles.	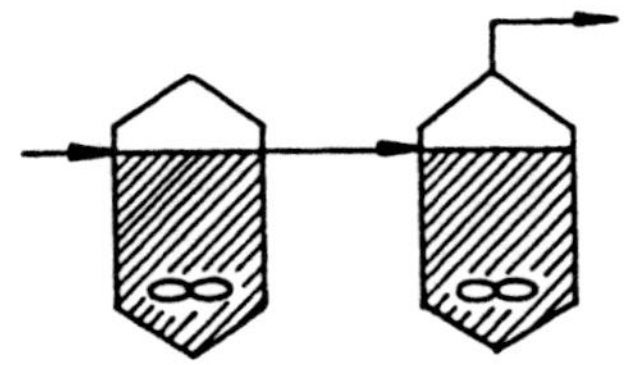

0.6.2. Continuous systems without accumulation of active biomass

An alternative to the batch digester is the completely mixed continuous system (Table 0.1, system 3). Here fresh substrate biomass is added, and digested mixed liquor is removed regularly, if not continuously. The biogas production rate remains more or less constant in a constant volume methane digester running under stationary conditions. A mean average production rate of 1 m^3 of biogas per m^3 working volume of digester per day can normally be obtained.

Several digesters are operating on this principle, as can be observed in Part I, Chapter I. One may cite as an example digester 80 in Thorembais-Les-Beguines (Belgium).

The completely mixed system has nevertheless some limitations. A first limitation is that, for mechanical reasons, the concentration of the waste to be treated may not exceed 100 kg dry matter per m^3 of influent waste. At higher concentrations the wastes are no longer pumpable.

Furthermore, each time the methane digester is fed, the same quantity of digested mixed liquor comes out of the digester and carries away also part of the bacteria (active biomass). Moreover, the methanogenic bacteria have a low specific growth rate and require about two to five days to double their number. As a result, the active biomass remains limited in a completely mixed continuous system and this, in time, limits the maximum possible biogas production rate (Nyns, 1979). Besides, in completely mixed continuous systems, the mean solid detention time (detention time of the solid part of the sludge containing the active biomass) is the same as the mean hydraulic detention time (detention time of the liquid part of the sludge). As a result, the mean hydraulic detention time may not be lowered below 10 days and the daily loading rate cannot exceed 4 kg of dry organic matter introduced in the methane digester per m^3 working volume of the digester per day. At higher loading rates and/or shorter mean detention times, intermediary volatile fatty acids will accumulate in the methane digester and hamper methane production.

Possible variants of the completely mixed continuous system are the partially mixed continuous system and the plug-flow continuous system depicted in Table 0.1, items 4 and 5. Digester 87 in Braunschweig (F.R. Germany) described in Part I, Chapter I illustrates the partially mixed system whereas the plug-flow system may be illustrated by digester 6 located in Haslev (Denmark).

0.6.3. Continuous systems with accumulation of active biomass

In order to improve the completely mixed continuous system, the active biomass may be trapped and accumulated inside the methane digester. In this way, higher solid mean detention times than the hydraulic detention times can be obtained.

The first continuous system with accumulation of active biomass was the anaerobic contact process shown as item 6 in Table 0.1. The contact process was first assayed by Steffen and Bedker on meat packing wastes in 1962. In that case, after methane digestion in a completely mixed continuous system, the active biomass was separated from the effluent in a

Duiven (The Netherlands) 1979. 120 m^3 completely mixed continuous methane digester on dairy cattle manure.

Renescure (France) 1981. Twin 2 500 m^3 anaerobic contact methane digesters on vegetable canning wastewater.

Argelato (Italy) 1982. Twin 1 250 m^3 anaerobic filter methane digesters or sugary wastewater.

Breda (The Netherlands) 1980. 1 300 m^3 upflow without carrier methan digester on sugary wastewater.

settler. The mean hydraulic detention times could thereby be reduced from 5 days to 1/2 day which allowed a substantial reduction in the size of the biogas plant. Mean average gas production rates of 2 m^3 biogas per m^3 of digester working volume per day were obtained. In Part I, Chapter I, digester 534 located in Renescure (France) is a typical example of the anaerobic contact process.

The contact process is nevertheless rather difficult to control, essentially in the settling step. A variant of this process is the IRIS system, investigated by the Institut de Recherche de l'Industrie Sucrière located at Lille (France). The system was developed by Lescure and co-workers (Devillers et al., 1977) and is shown as item 7 in Table O.1. The settler is in this case incorporated within the methane digester and is merely separated from it by a tarpaulin or a steel wall. The active biomass falls by gravity inside the methane digester. Digester 487 of Lecoudray Montceau (France) (see Part I, Chapter I) is one of the IRIS digesters.

Another way to retain the microorganisms inside the methane digester is the anaerobic filter (Table O.1, item 9) first tried by Young and McCarty in 1969. This system consists in trapping and immobilizing the active biomass on fixed carriers such as stones, clay, ceramic or plastic pieces or matrices. Colleran uses, for instance, plastic rings of 0.94 porosity for her anaerobic filter running on pig manure (see digester 236 of Galway (Ireland) in Part I, Chapter I). Since the mean hydraulic detention times can be reduced to 2 - 3 days, the biogas production rate will increase up to 4 m^3 of biogas produced per m^3 of digester working volume per day. These anaerobic filters nevertheless present problems of clogging.

A third system for improving the performance of the methane digester by retaining the active biomass inside the digester is the upflow anaerobic sludge blanket system, known as the UASB process, developed by Lettinga et al. in 1975, and shown in Table O.1, item 8 as upflow without carrier. The property of flocculation of the anaerobic active biomass is exploited in this system and leads to the accumulation of non-fixed bacteria in heterogeneous flocs. The settler/gas separator, placed in the upper part of the reactor, maintains the active biomass inside the latter. After settling, the active biomass forms a sludge bed in the lower part of the methane digester.

The technology of the UASB has already been demonstrated on several types of agro-industrial wastewaters, especially in The Netherlands. As an example, digester 506 in Koog a/a Zaan, treating starch factory wastewaters is proposed in Part I, Chapter I. By employing this system, mean hydraulic detention times as short as 6 hours can be achieved. But, because the biological flocs are not protected, this system remains susceptible to load variations and to the presence of toxic compounds in the influent.

A last system consists in immobilizing the active biomass either on a very porous matrix or on micro-carriers, i.e. inert solid grains of high density and small diameter. This leads either to a fixed film process or to the fluidized or expanded bed process shown in Table O.1, item 10. In the latter system, granules constituted by the active biomass immobilized on the mobile micro-carriers are fluidized in the major portion of the reactor by application of high upward liquid superficial velocities which are generated and can be regulated by liquid recycle (Switzenbaum and Jewell, 1980; Binot et al., 1983). In Finland, there exists already a full-scale fluidized bed treating kraft pulp bleaching effluents (Salkinoja-Salonen et al., 1982).

0.6.4. Two-step systems

The biology of the biomethanation process can be considered as a liquefaction-fermentation step, followed by a methanogenic step (Fig. 0.4). The two-step biomethanation system is based on the separation of the two biological steps. The system, depicted in Table 0.1, item 11, is characterized by two separate reactors, each of them optimized for one of the two biological steps. Each reactor can be based on any of the principles described in Table 0.1 for single-step systems. By separating the mean hydraulic detention times and the loading rates, corresponding to the two biological steps, improved gas yields and rates are claimed. Indeed, the supply of a liquefied effluent to a subsequent methane digester might allow a better process control. Overloading and consequent "souring" of the methane digester can be prevented. Furthermore, by raising the turnover of the whole process, the overall plant size can be reduced. Digester 204 located in Carobbio (Italy) (see Part I, Chapter I) is an example of a two-step system, based on the completely mixed principle.

0.6.5. Landfills

Special mention must be made of a biomethanation system particularly adapted to domestic residues or urban refuse. In all European countries a large part of the domestic refuse has been and still is disposed of on landfill sites. In fact, these sites can be considered as large-scale methane digesters, since they present favourable conditions for producing biogas, namely the presence of organic matter, the absence of air, high temperature originating from a primary (aerobic) decomposition of the organic matter, and long-time storage.

The production of biogas from landfills is a natural phenomenon, and explosions often occurred. This led to gas recovery, originally for security reasons. Biogas recovery is at present mainly realized for economic reasons. Since 1979, test sites and full-scale demonstrations have increased significantly in more than half of the European countries, as will appear from the survey.

In order to abstract the biogas from the landfill, pits are sunk at the site. These pits contain pipes which are perforated at their lower part and are surrounded by filtering material. Fig. 0.5 depicts one of these pits. After abstraction, the biogas is purified and used either on the landfill site itself or in a nearby industry or farm. In the brick - landfill of London, for example, the biogas is used in the ovens for baking bricks.

Biogas recovery from landfills can be improved in two ways. First, the liquid resulting from the percolation of water (e.g. rain) through the landfill can be recovered at the bottom part of the landfill and recycled. This leads to a more appropriate humidity within the landfill and promotes bacterial activity. The addition of sewage sludge to the domestic refuse

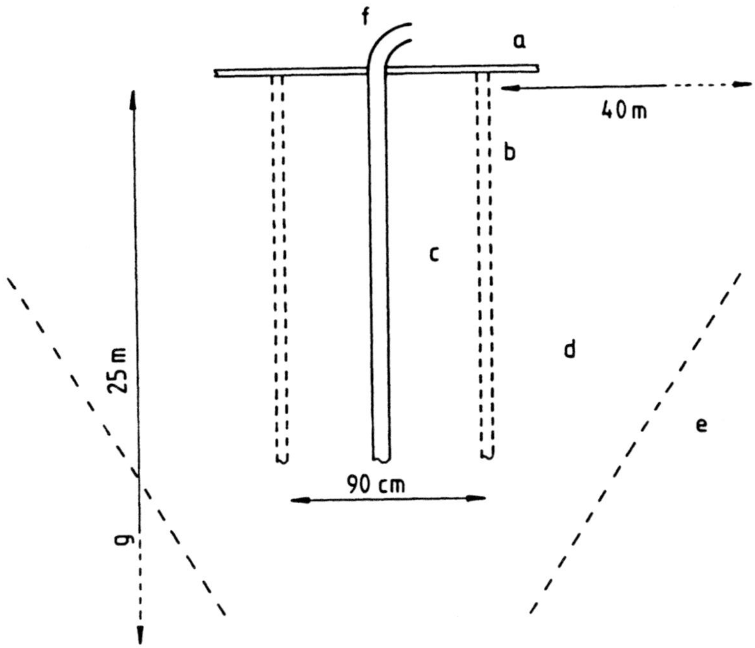

Fig. 0.5. Pit for biogas abstraction in landfills

a : Gas-proof cover; b : Limit between gravel and organic matter; c : Gravel filling; d : Worked area; e : Non-influenced area; f : Perforated pipe in polyvinyl chloride (PVC); g : Depth of the pit : 60 to 80 % of the total thickness of the organic matter layer.

in the landfill site is also claimed to improve the biogas productivities (Tabasaran, 1982).

Secondly, when the site is still on filling, it can be divided into cells by clay walls. The brick-landfill of Stewartby near London is divided in this way (see Part I, Chapter I, Digester 456). Abstraction of biogas from these cells can thus be better controlled since interference of one pit with another is avoided.

0.7. Wastes versus biomethanation systems

Depending upon the physical state of the wastes treatable by biomethanation, a minimum of three categories may be distinguished. First are the solid wastes having more than 10 % dry matter concentration (wt wt^{-1}), including the solid agricultural wastes such as manure with bedding or crop residues, the organic part of domestic refuse and the solid agro-industrial wastes such as paunch manure or stomach content. To this category belong also the energy crops.

The second category comprises the semi-solid wastes or slurries which are basically pumpable; their dry matter concentration exceeds 4 % (wt wt^{-1}). They include liquid wastes with a high content of suspended solids and essentialy all types of manure produced on farms. Third are the liquid wastes, characterized by low concentrations of suspended solids, including wastewaters of agro-industrial origin. Their dry matter concentration does not usually exceed 4 % (wt wt^{-1}).

As regards the applicability of the different biomethanation systems to these three categories of wastes, the batch system can be used, in principle, for all three categories, the completely mixed continuous system can be used for the last two categories, and all systems with recycle or trapping of active biomass are applicable to the liquid wastes only.

Nevertheless, these ranges of applicability can be extended when the waste is pretreated before being biomethanized. Solid wastes as such can only be biomethanized in batch systems. But when the solid waste is diluted and ground the resulting mixture can be treated as a slurry. Many examples of completely mixed or plug-flow systems running on solid wastes were encountered in the survey, although this procedure is not necessarily the most economical way to biomethanize solid wastes.

In some cases, such as that of non-lignified energy crops, the solid waste can also be biologically liquefied to a slurry. A two-step process using two completely mixed continuous reactors is then appropriate. A third possibility consists in a preliminary biological percolation. In this pretreatment, water or a liquid is recirculated through a heap of solid wastes and the liquid phase resulting from the percolation is then further submitted to biomethanation. For the latter step, the more advanced biomethanation system with active biomass entrapment is used. Only pilot-scale systems based on the former principle were encountered in the present survey.

When slurries are pretreated by percolation or simply diluted, they can also be biomethanized by the more advanced biomethanation systems with active biomass entrapment. The best example of this is the biomethanation of pig manure on anaerobic filters. Pig manure is already well diluted when it is produced. A simple clarification step suffices to produce a liquid which can be biomethanized in an anaerobic filter. After successful pilot-scale demonstrations both in Ireland and France, a full-scale anaerobic filter on pig manure is under construction in France and is planned to be in operation early in 1984.

0.8. Potentialities of biomethanation for energy production in Europe

Several available or potential biomass resources, not only wastes and residues but also energy crops, can be biomethanized and, in this way, will contribute to the energy requirements of the European Community. The energy potential of agricultural wastes and biomasses in Europe has recently been reviewed (Palz and Chartier, 1980). The potential contribution from available and potential resources to the energy requirements of the European Community, in the form of biogas, are summarized in Table 0.2.

Table 0.2. Potential contributions from available and potential resources to the energy requirements of the European Community

Categories of wastes and crops	Energy available as biogas (Mtoe $year^{-1}$)
- Accessible animal wastes	14.1 Mtoe (1)
- Crop residues	8.5 Mtoe (1)
- Municipal solid wastes	8.9 Mtoe (2)
	31.5 Mtoe
- Sewage sludge	1.8 Mtoe (2)
- Catch crops, algae and aquatic plants	8.5 (2) - 57 (1) Mtoe

(1) Palz and Chartier (1980)
(2) Hall (1983)

Table 0.2 shows that the conversion of available animal wastes, crop residues and municipal solid wastes can produce yearly an amount of usable energy approximating to 30 million ton oil equivalent. This figure looks small because it represents only about 3 % of the estimated European energy consumption by 1985 (Hall, 1983). It represents however a potential yearly market value of about 7 000 million ECU.

The biomass resources for which biomethanation is the appropriate energy-converting technology, as indicated in Table 0.2, are the wastes containing over 35 % moisture. A first category comprises the so-called wet agricultural wastes, crop residues and municipal solid wastes. As a point of reference, all European sewage sludges in wastewater treatment plants amount to only about 6 % of this first category of potential substrates for biomethanation (Hall, 1983). A small portion of such wet resources are already currently treated by biomethanation.

As well as wastes and residues one should take into account energy crops, i.e. plants grown for the sole or main purpose of energy production, e.g. catch crops, algae or other aquatic plants. The potential of these energy crops for biomethanation, non-competitive with food crops, has been estimated as high as twice the value for wastes and residues (Table 0.2). Water hyacinths are already currently used at the same time for wastewater treatment and biogas production, namely at Disney World at Orlando, Florida, U.S.A. (Lee and McKim, 1981).

However, this second category of biomass resources is not yet used for biomethanation in Europe. Research and development is being promoted and will lead in the mid-term future to demonstration biogas plants in the southern part of Europe. Project E, "Energy from Biomass", of the Energy Research and Development Programme of the Commission of the European Communities, aims at the introduction of energy crops for biofuels production by the year 2000.

PART I : TECHNOLOGIES AND ECONOMICS OF BIOGAS PLANTS

CHAPTER I. FLOWSHEETS OF TWENTY-SEVEN REPRESENTATIVE BIOGAS PLANTS

In the following pages, 27 selected biogas plants are described. Of these 27 plants, 17 are located on farms, 6 are in industry, 1 is a landfill, and 3 are pilot-scale biogas plants.

The selection was based on the following criteria :

- the methane digesters must be operating satisfactorily and be well monitored, with sufficient data available,
- they must differ either in the type of feedstock treated or the methane digestion system used, or in running conditions.

Each of the 27 biogas plants is identified by its town and initial of country and by a number referring to the listing presented in Appendix III.

A draft depicts each biogas plant. To facilitate understanding of the drafts, a common legend is given in Fig. I.1.1.

Each draft gives the address, the start-up and the cost of the biogas plant. The design of the methane digester is explained by indicating its biomethanation system, digestion working volume, the material used for construction, the type of insulation, and heating and mixing devices. Details are also given of the origin of the feedstock and its treatment before biomethanation. The running conditions detail the volumetric loading rate (given either as to volatile solids (VS), or total solids (TS) or as to chemical oxygen demand (COD)), the temperature at which methane digestion is performed, and the mean hydraulic or solids detention times. Results on gas productivities -the rates, yields and gas composition- are reported, as well as the observed conversion which corresponds to the depollution attained. The storage device and the uses of the biogas produced are indicated for each plant. The treatment and the uses of the effluent are mentioned. Finally, the major problems encountered with each biogas plant are listed.

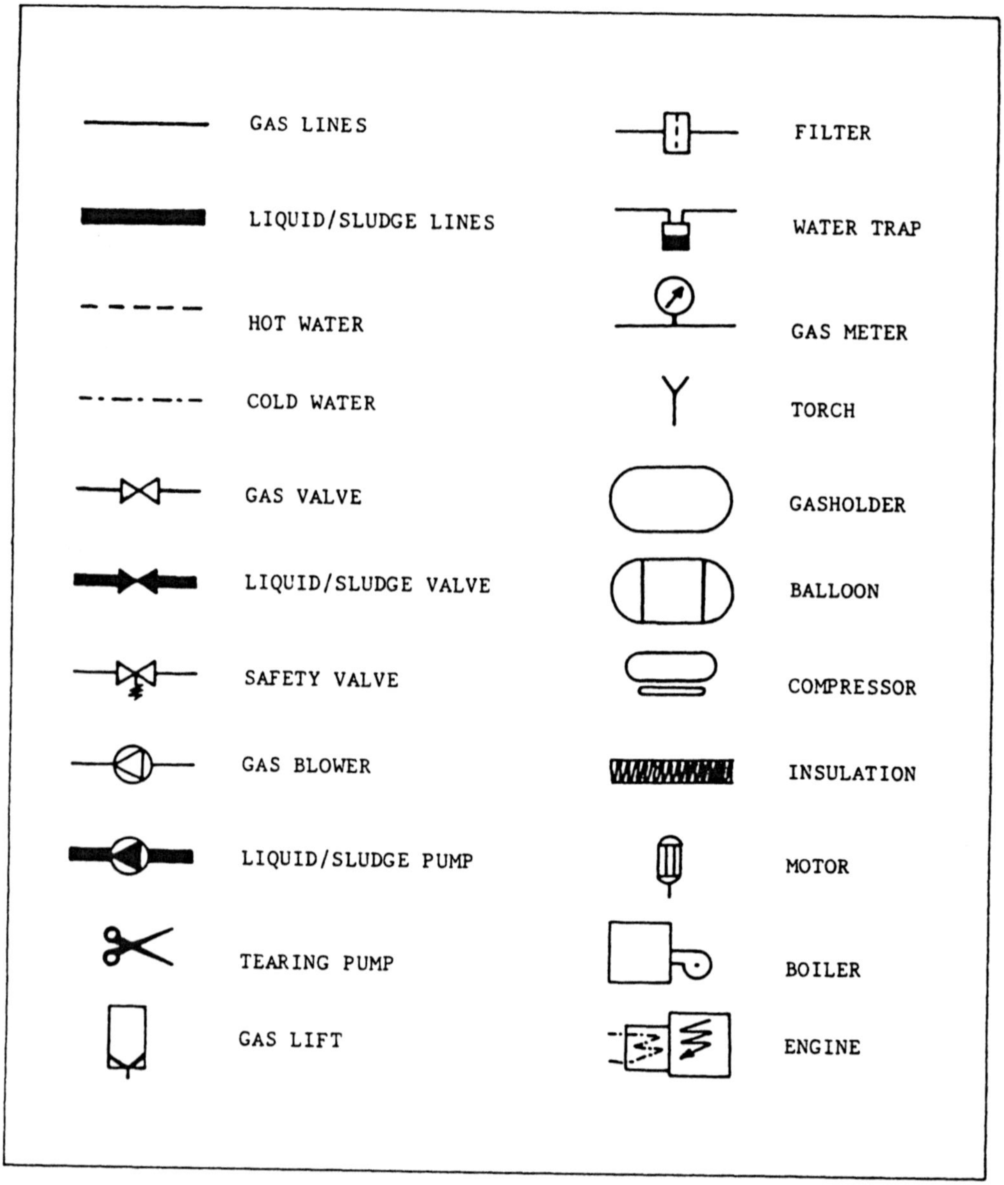

Fig. I.1.1. Common legend of the drafts of the 27 biogas plants

F-MARCHESIEUX - DIGESTER 315

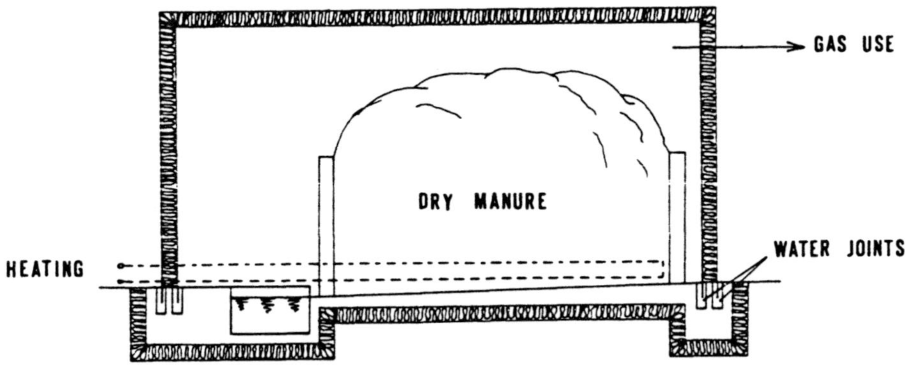

ADDRESS : Hameau de la Bizotière
F-50190 MARCHESIEUX, France

START-UP : 03.1981

DESIGN OF THE DIGESTER :

- System : discontinuous (batch); unimmersed
- Volume : total : 47 m^3, working volume = 24.5 m^3
- Material : concrete (platform), reinforced polyester tilted cover
- Insulation : expanded polystyrene
- Heating : external underground serpentine
- Mixing : liquid recirculation (pump)

COST : -

FEEDSTOCK : Dry bovine cattle manure with bedding (the whole year)

TREATMENT : Natural aerobic prefermentation

RUNNING CONDITIONS :

B_V : 3 kg VS m^{-3} d^{-1}

T = 34 $\pm$ 5 °C

θ_H = 56 d

θ_S = 56 d

GAS :

- Storage : floating gasholder
- Production : rate : 0.5 m^3 m^{-3} d^{-1}
  : yield : 0.17 m^3 kg^{-1} VS_o
- Composition : 54 % CH_4
- Depollution : 36 % VS reduction
- Uses : 18.5 % internal use (June-July)

EFFLUENT :

- Treatment : none
- Uses : soil conditioner

MAJOR PROBLEMS :

None for the moment

CH-DREIEN - DIGESTER 256

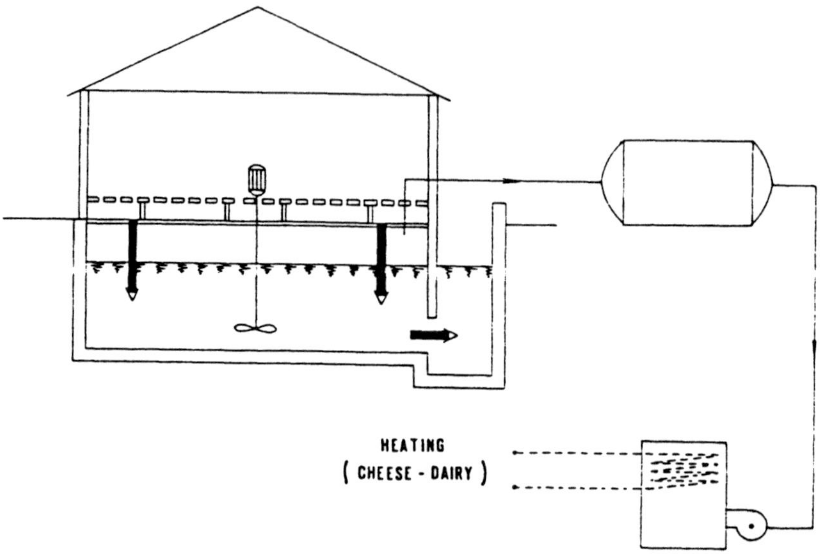

ADDRESS : M. H. Güntensperger
CH-9601 DREIEN, Switzerland
START-UP : 11.1979
DESIGN OF THE DIGESTER :
- System : fed-batch,
- Volume : total = 300 m^3
- Material : concrete
- Insulation : polymeric (6 cm)
- Heating : none
- Mixing : mechanical (agitation axle + motor)
COST : SFR 27 000 (without labour cost of the farmer)

FEEDSTOCK : Manure from 350-400 fattening pigs fed with mixed grain and whey
TREATMENT : None
RUNNING CONDITIONS
B_V = -
T = 15-20 °C
θ_H = -
θ_S = -

GAS :
- Storage : balloon
- Production : rate : 0.20 $m^3\ d^{-1}\ m^{-3}$
: yield : -
- Composition : 61 % CH_4, 39 % CO_2
- Depollution : -
- Uses : heating of the cheese dairy
EFFLUENT :
- Treatment : none
- Uses : soil conditioner
MAJOR PROBLEMS :
None for the moment, unless very slow start-up

B-THOREMBAIS-LES-BEGUINES - DIGESTER 80

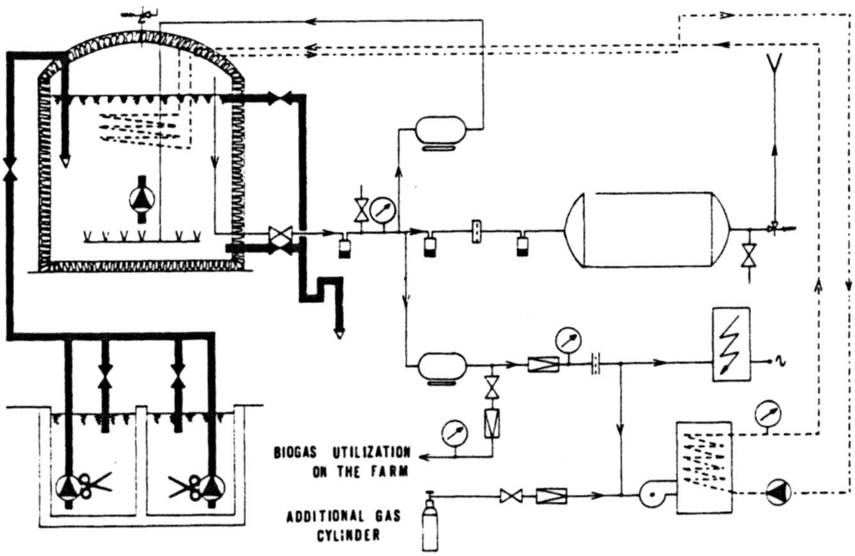

ADDRESS : M.L. Majoie, chaussée de Charleroi, 5
B-5921 THOREMBAIS-LES-BEGUINES, Belgium

START-UP : 02.1982

DESIGN OF THE DIGESTER :

- System : completely mixed digester, intermittent loading
- Volume : total : 136 m^3, working volume : 100 m^3
- Material : double wall in steel
- Insulation : type of mica vermiculite + mineral wool
- Heating : internal serpentine
- Mixing : internal centrifugal pump + possible gas recirculation

COST : 2 200 000 BFR (total)

FEEDSTOCK : Manure from 90 beef cattle from 3 months to 2 years, during the whole year on grating

TREATMENT : Storage of one month + dilution + grinding

RUNNING CONDITIONS :

$B_V = 1.6$ kg VS m^{-3} d^{-1}

$T = 35 \pm 1$ °C

$\theta_H = 31$ d

$\theta_S = 31$ d

GAS :

- Storage : balloon in hypalon (50 m^3)
- Production : rate : 0.60 m^3 d^{-1} m^{-3}
  : yield : 0.38 m^3 kg^{-1} VS_o
- Composition : 61 % CH_4
- Depollution : 40 % VS reduction
- Uses : ± 40 % for the digester heating, no other use for the moment

EFFLUENT :

- Treatment : none
- Uses : soil conditioner

MAJOR PROBLEMS :

None for the moment

B-VOORMEZELE - DIGESTER 145

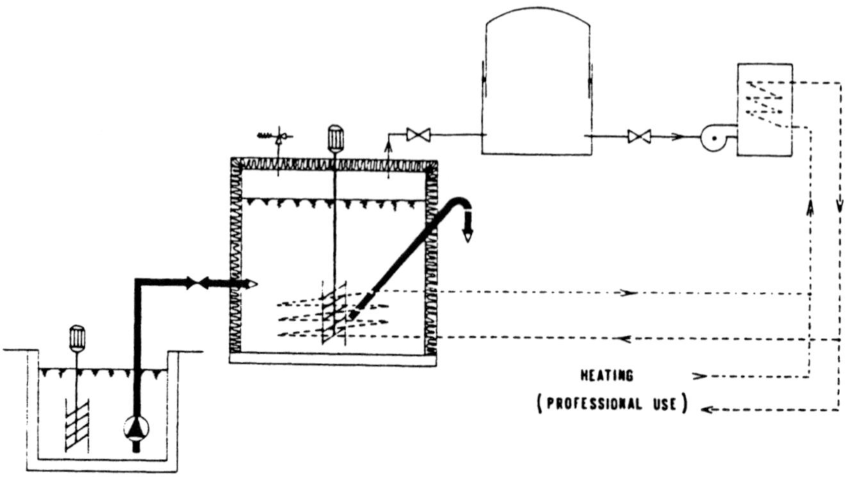

ADDRESS: M. Decapmaker, St Elooiweg, 47
B-8903 VOORMEZELE, Belgium

START-UP : 03.1981

DESIGN OF THE DIGESTER :

- System : completely mixed digester, intermittent loading, one step
- Volume : total = 87 m^3, working volume = 73 m^3
- Material : concrete (int. coating : polyurethane, ext. coating : plastering + insulation)
- Insulation : polystyrene (6 cm)
- Heating : internal serpentine
- Mixing : mechanical screw (motor)

COST : 400 000 BFR

FEEDSTOCK : Manure from 50 sows and 350 fattening pigs fed with compounds of barley, soya and maize during the whole year

TREATMENT : Storage one week to one month

RUNNING CONDITIONS :

B_V = 2-3 kg COD m^{-3} d^{-1}
T = 30 °C
θ_H = 35 d
θ_S = 35 d

GAS :

- Storage : floating gasholder (14 m^3) in Alu. protected by a polyurethane internal & external envelope
- Production : rate : 0.48-0.72 m^3 m^{-3} d^{-1}
  : yield : 0.24 m^3 kg^{-1} COD_o
- Composition : 65-70 % CH_4
- Depollution : -
- Uses : 30 % internal use + 70 % professional use : heating

EFFLUENT :

- Treatment : none
- Uses : soil conditioner

MAJOR PROBLEMS :

None for the moment

DK-GRAASTEN - DIGESTER 338

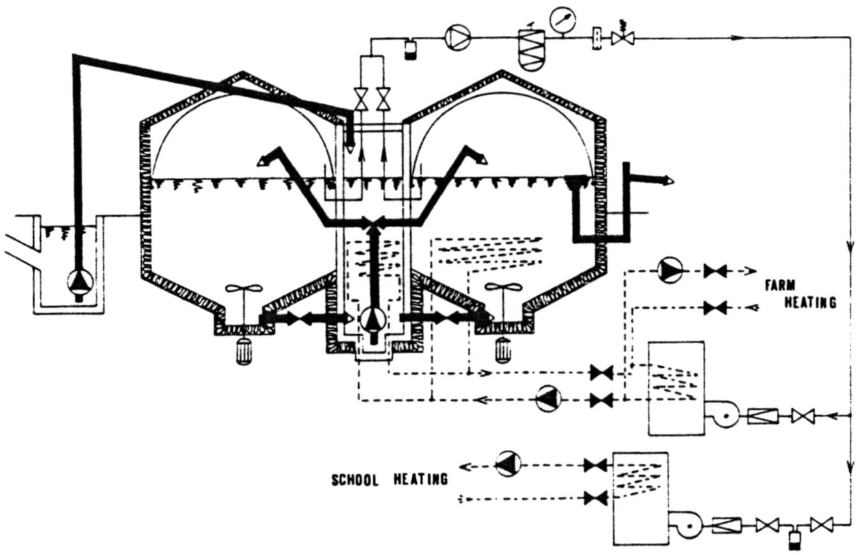

<u>ADDRESS</u> : Fiskebaekgaard, Graasten Lanbrugsskole
DK-6300 GRAASTEN, Denmark

<u>START-UP</u> : 09.1979

<u>DESIGN OF THE DIGESTER</u> :

- System : 2 completely mixed digesters in series, intermittent loading
- Volume : total = 2 x 220 m^3 + gas balloons, working volume 2 x 190 m^3
- Material : concrete
- Insulation : mineral wool (15 cm)
- Heating : 20 m^3 external heating chamber + serpentine (mild steel)
- Mixing : liquid recirculation : pump + impeller in the centre bottom

<u>COST</u> : 750 000 DKR (+ 22 % VAT)

<u>FEEDSTOCK</u> : Manure from 96 dairy cattle + 500 porkers + 20 sows the whole year + 50 calves in Winter fed with raw fodder and silage (cows), compounds (pigs)

<u>TREATMENT</u> : Storage 1-2 days + dilution + grinding (cutting of straw by knives mounted in the feed pump)

<u>RUNNING CONDITIONS</u> :

B_V = 2.5 kg VS m^{-3} d^{-1}

T = 35 $\pm$ 3 °C

Θ_H = 25 d

Θ_S = 25 d

<u>GAS</u> :

- Storage : balloon in PVC (papyrotex) 1 mm
- Production : rate : 0.8-0.92 m^3 d^{-1} m^{-3}
  : yield : 0.35-0.41 m^3 kg^{-1} VS_o
- Composition : 58-65 % CH_4
- Depollution : -
- Uses : 33 % internal use + 67 % heating (domestic use)

<u>EFFLUENT</u> :

- Treatment : decantation + filtration
- Uses : soil conditioner

<u>MAJOR PROBLEMS</u> :

- Scum formation, inefficient mixing device before impellers were installed

D-BOCKENEM - DIGESTER 150

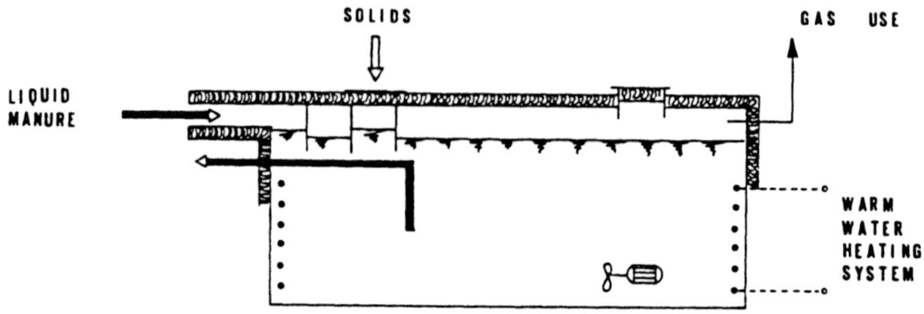

ADDRESS : Fischer Heinz, Glashütter Strasse, 26
D-3205 BOCKENEM 3, Federal Republic of Germany
START-UP : 08.1981
DESIGN OF THE DIGESTER :
- System : completely mixed, intermittent loading
- Volume : 280 m^3
- Material : concrete
- Insulation : styrodur
- Heating : internal serpentine
- Mixing : internal propeller stirrer
COST : 35 000 DM

FEEDSTOCK : Manure from 74 livestock units breeding and fattening pigs
TREATMENT : Storage of 0 to 7 days in gravity channels
RUNNING CONDITIONS :
B_V = -
T = 10-15 °C
θ_H : 90 d
θ_S : > 90 d

GAS :
- Storage : 18 m^3 rubber balloon (15 mbar)
- Production : rate : 0.1 $m^3\ d^{-1}\ m^{-3}$ (increasing to 0.2 $m^3\ d^{-1}\ m^{-3}$ in 1983
: yield : -
- Composition : 50 % CH_4
- Depollution : -
- Uses : 100 % heating (professional use)
EFFLUENT :
- Treatment : none
- Uses : soil conditioner
MAJOR PROBLEMS :
None for the moment

D-WEDEMARK - DIGESTER 407

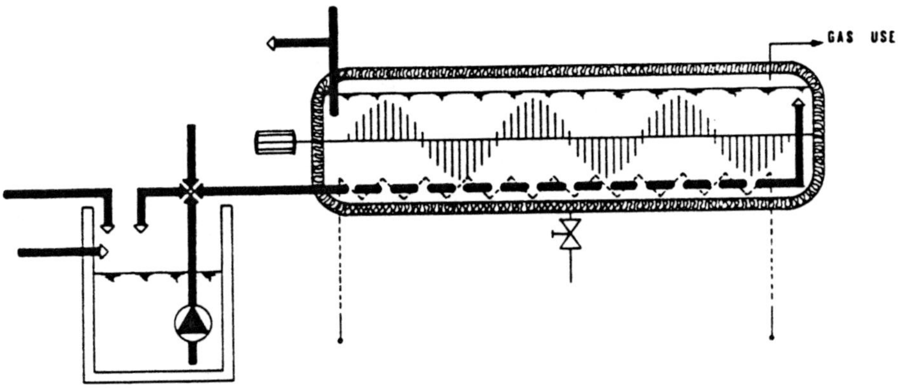

ADDRESS : Hemme Hermann, Marstrasse, 3
D-3002 WEDEMARK 1, Federal Republic of Germany

START-UP : 06.1981

DESIGN OF THE DIGESTER :
- System : horizontal through-flow, completely mixed, intermittent loading
- Volume : 60 m^3
- Material : steel
- Insulation : mineral wool (20 cm)
- Heating : internal serpentine
- Mixing : horizontal mechanical axle, motor

COST : 60 161 DM

FEEDSTOCK : Manure from 27 livestock units fattening pigs + 22 livestock units fattening bulls + straw and silage (± 30 kg TS/d)

TREATMENT : Storage of 7 to 14 days in gravity channels

RUNNING CONDITIONS :
B_V = 5 kg VS m^{-3} d^{-1}
T = 33-38 °C
θ_H = 22 d
θ_S = > 22 d

GAS :
- Storage : none
- Production : rate : 1.1 m^3 d^{-1} m^{-3}
  : yield : 0.49 m^3 kg^{-1} VS_o
- Composition : 60 % CH_4
- Depollution : -
- Uses : 25 % internal use + 60 % heating (professional use)

EFFLUENT :
- Treatment : none
- Uses : soil conditioner

MAJOR PROBLEMS :
None for the moment

I-VEDELAGO - DIGESTER 232

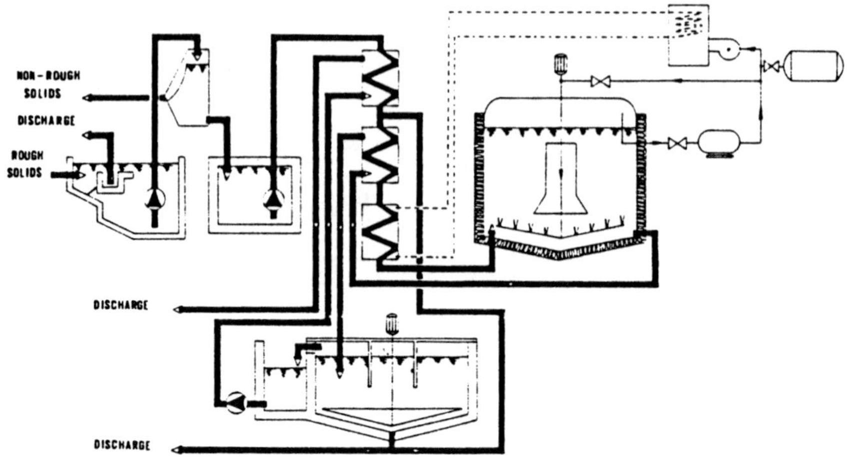

ADDRESS : Cooperativa San Pio X
I-VEDELAGO (Treviso), Italy

START-UP : 12.1980

DESIGN OF THE DIGESTER :

- System : continuous, completely mixed digester
- Volume : total = 930 m^3, working volume = 780 m^3
- Material : concrete, int. coating : epoxidic paint, ext. coating : zincate steel sheet
- Insulation : polymeric (5 cm)
- Heating : external heat exchanger
- Mixing : gas recirculation + mechanical (agitation axle)

COST : 350 000 000 LIT

FEEDSTOCK : Manure from 7 000 swines the whole year + 5 000 swines in winter and 5 000 swines in summer, all fed with whey and silage (+ Cu + antibiotics)

TREATMENT : Storage 15 h + static screening

RUNNING CONDITIONS :

B_V = 1.79 kg VS m^{-3} d^{-1}
T = 35 °C
θ_H = 4 d
θ_S = 25 d

GAS :

- Storage : none
- Production : rate : 1.0 m^3 d^{-1} m^{-3}
  : yield : 0.56 m^3 kg^{-1} VS_o
- Composition : 70 % CH_4
- Depollution : 79 % VS reduction
- Uses : 43 % internal use + 57 % heating (professional use)

EFFLUENT :

- Treatment : 500 m^3 activated sludge tank
- Uses : discharge in surface water

MAJOR PROBLEMS :

Settlement during storage, build-up of deposit on heat exchanger surface causing ineffective heat transfer, gas leakage, freezing H_2O in line during winter operation, CH_4 diffusion through pipes on storage unit, electrical failure, gas monitoring devices

NL-DUIVEN - DIGESTER 28

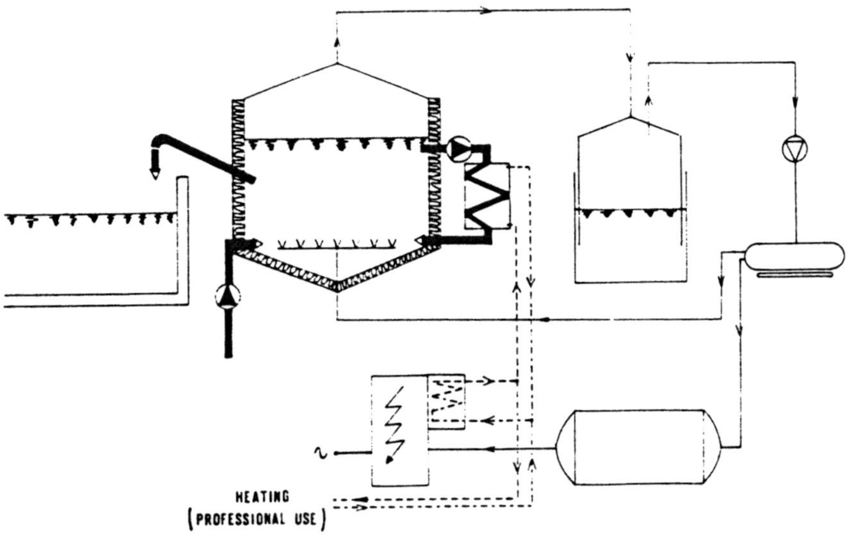

<u>ADDRESS</u> : Experimental dairy farm "De Vijf Roeden", Zesroedenweg, 1
NL-DUIVEN, The Netherlands

<u>START-UP</u> : 11.1979

<u>DESIGN OF THE DIGESTER</u> :
- System : completely mixed digester, intermittent loading
- Volume : total = 120 m^3; working volume = 104 m^3
- Material : steel enamelled both sides
- Insulation : polymeric (6 cm)
- Heating : external heat exchanger
- Mixing : gas and liquid recirculation (pump + agitation axle)

<u>COST</u> : 160 000 HFL

<u>FEEDSTOCK</u> : Manure (6 m^3 d^{-1} in winter, 2 m^3 d^{-1} in summer) (8 % TS) from 80 dairy cattle + 40 calves, fed with power fodder during the whole year, with grass in summer, and silage (maize + + grass) in winter

<u>TREATMENT</u> : Storage 1-2 months + dilution (± 20 %) + grinding

<u>RUNNING CONDITIONS</u> :
B_V = 1.8 kg VS m^{-3} d^{-1}
T = 30 °C
θ_H = 39.5 d
θ_S = 39.5 d

<u>GAS</u> :
- Storage : small floating gasholder + compressor + high pressure tank in mild steel
- Production : rate : 0.36 m^3 d^{-1} m^{-3}
  : yield : 0.21 m^3 kg^{-1} VS_o
- Composition : 64.3 % CH_4
- Depollution : 17 % VS reduction
- Uses : 56 % internal use + 10 % electric power + hot water for professional use

<u>EFFLUENT</u> :
- Treatment : none
- Uses : soil conditioner

<u>MAJOR PROBLEMS</u> :
Failure of pumping device, operational problems with heating of feed, failure or inefficient heat transfer, compressor problems during gas pumping, electricity generation engine problems

NL-KLARENBEEK - DIGESTER 240

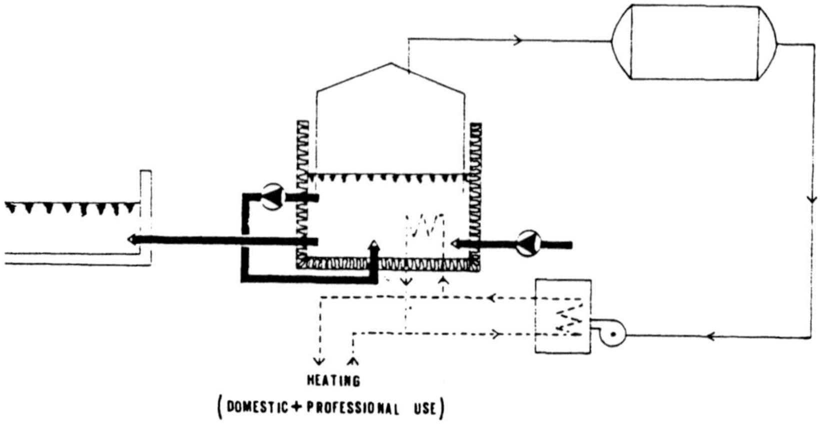

ADDRESS : H. Hobert, Koperdijk, 5
NL-KLARENBEEK, The Netherlands

START-UP : 12.1979

DESIGN OF THE DIGESTER :

- System : completely mixed digester, intermittent loading
- Volume : total = 115 m^3, working volume = 75 m^3
- Material : steel enamelled both sides
- Insulation : polymeric (6 cm)
- Heating : internal serpentine
- Mixing : liquid recirculation (pump)

COST : 68 000 HFL

FEEDSTOCK : Manure from 1000 swines fed with compounds (+ Cu) the whole year (+ antibiotics only in case of animal disease)

TREATMENT : Storage up to 2.5 months + dilution

RUNNING CONDITIONS :

B_V = 2.4-2.7 kg VS d^{-1} m^{-3}

T = 30 °C

θ_H = 21-25 d

θ_S = 21-25 d

GAS :

- Storage : balloon (polymeric) + floating gas holder inside the digester
- Production : rate : 0.67-0.80 m^3 d^{-1} m^{-3}
  : yield : 0.30 m^3 kg^{-1} VS_o
- Composition : 67.8 % CH_4
- Depollution : 20-25 % TS reduction
- Uses : 35 % internal use + heating (32.5 % domestic use + 32.5 % professional use)

EFFLUENT :

- Treatment : none
- Uses : soil conditioner

MAJOR PROBLEMS :

- Digester too small
- Pump leakage

UK-EDENBRIDGE - DIGESTER 37

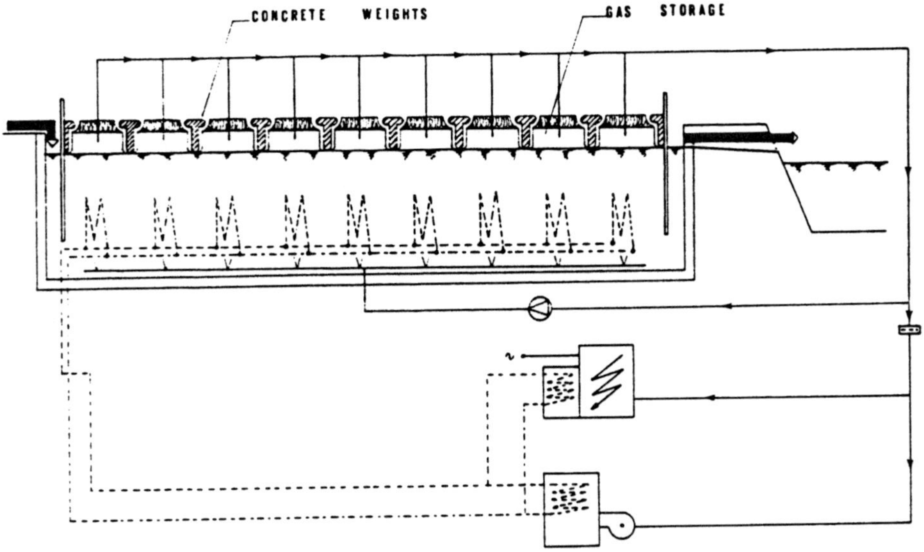

ADDRESS : N.E. Wates, Bore Place, Chiddingstone
UK-EDENBRIDGE (Kent), United Kingdom

START-UP : 10.1979

DESIGN OF THE DIGESTER :

- System : completely mixed digester, continuous loading
- Volume : total = 324 m^3; working volume = 300 m^3
- Material : concrete
- Insulation : polymeric insulation on cap (10 cm) + surrounding soil
- Heating : internal heat exchanger
- Mixing : gas recirculation (pump)

COST : UKL 86 000

FEEDSTOCK : Manure from 320 dairy cattle fed with compounds and silage in winter

TREATMENT : Dilution

RUNNING CONDITIONS :

B_V : 5.57 kg VS.$d^{-1}.m^{-3}$
T = 34.1-40 °C
Θ_H = 15.3 d
Θ_S = 15.3 d

GAS :

- Storage : floating gasholder (glass reinforced plastic)
- Production : rate : 0.77 m^3 d^{-1} m^{-3}
  : yield : 0.137 m^3 kg^{-1} VS_o
- Composition : 56 % CH_4
- Depollution : 33 % VS reduction
- Uses : professional electric power (dairy power)

EFFLUENT :

- Treatment : storage : earth-walled lagoon
- Uses : soil conditioner and fertilizer

MAJOR PROBLEMS :

None for the moment

CH-KALTBRUNN - DIGESTER 264

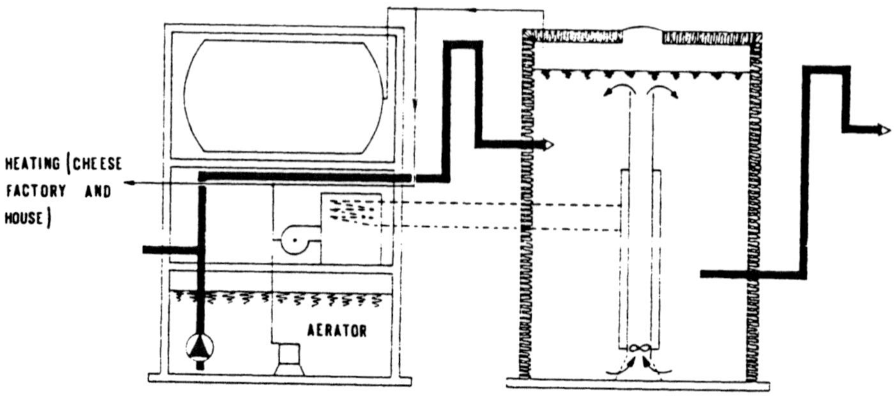

ADDRESS : E. Huber, Käserei Steinenbrugg
CH-8722 KALTBRUNN, Switzerland

START-UP : 06.1981

DESIGN OF THE DIGESTER :

- System : completely mixed; intermittent loading
- Volume : total = 157 m^3, working volume : 147.5 m^3
- Material : double wall, concrete
- Insulation : styrofoam (10 cm) + roofmate
- Heating : double tube heat exchanger (mixing tube)
- Mixing : internal mixing tube

COST : 100 000 SFR

FEEDSTOCK : Manure from 600 finishing hogs

TREATMENT : Storage + aeration

RUNNING CONDITIONS :

B_V = 1.78 kg VS m^{-3} d^{-1}
T = 36 °C
θ_H = 42 d
θ_S = -

GAS :

- Storage : balloon (110 m^3)
- Production : rate : 0.957 m^3 m^{-3} d^{-1}
  : yield : 0.538 m^3 kg^{-1} VS_o
- Composition : 70 % CH_4
- Depollution : -
- Uses : heating digester : 25 %, heating cheese factory and house : 75 %

EFFLUENTS :

- Treatment : not available
- Uses : not available

MAJOR PROBLEMS :

Not available

I-FAENZA - DIGESTER 492

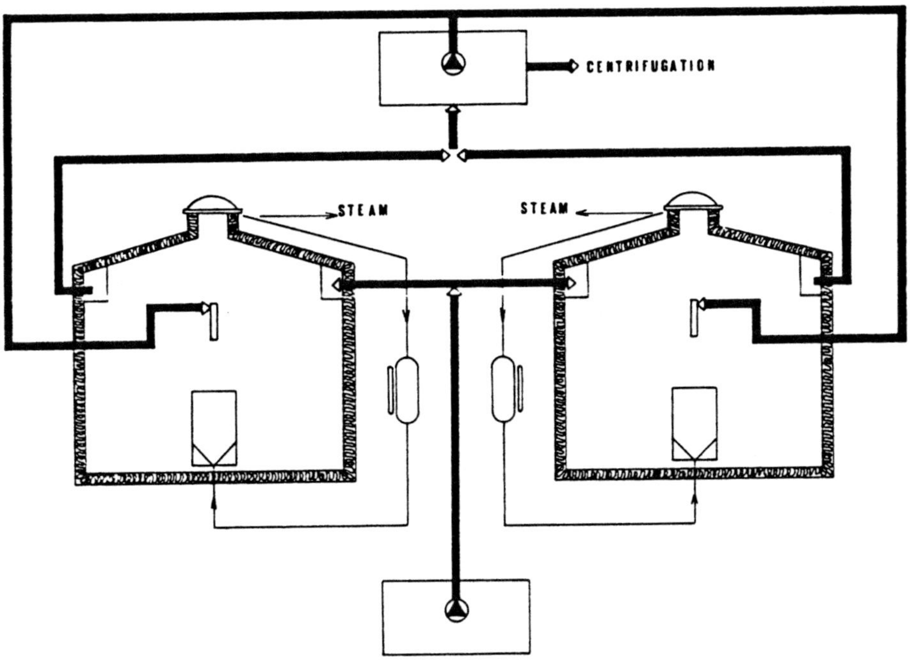

ADDRESS : NERI, Distillery - FAENZA (Ravenna) Italy
START-UP : April 1981
DESIGN OF THE DIGESTER :
- System : completely mixed, continuous loading
- Volume : total = 2 x 7 000 m^3, working volume = 2 x 6 000 m^3
- Material : steel
- Insulation : expanded polyurethane
- Mixing system : gas lift
- Heating system : not necessary since T of feedstock about 40 °C

COST : Not available

FEEDSTOCK : Distillery wastewaters (dreg, wine, fruits)
TREATMENT : Cooling to 40 °C, neutralization...
RUNNING CONDITIONS :
B_V = 2.7 kg VS m^{-3} d^{-1}
T = 35 °C
θ_H = 10 d
θ_S = 10-25 d

GAS :
- Storage : gasometer (500 m^3)
- Production : rate : 1.5 m^3 d^{-1} m^{-3}
: yield : 0.55 m^3 kg^{-1} VS_o
- Composition : 65 % CH_4
- Depollution : 55 % VS reduction
- Uses : 100 % professional use (steam)

MAJOR PROBLEMS :
Settling of digested mixed liquor resolved by centrifugation

UK-ASHFORD - DIGESTER 508

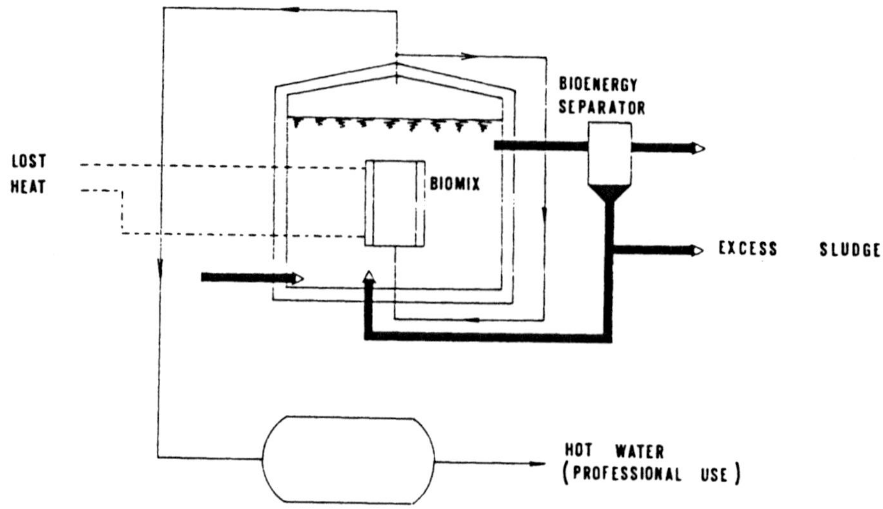

ADDRESS : Tenstar Product Ltd., Hithe Road
UK-ASHFORD (Kent), United Kingdom

START-UP : 1976

DESIGN OF THE DIGESTER :
- System : completely mixed digester, continuous loading
- Volume : total = 1 000 m^3; working volume = 900 m^3
- Material : steel, epoxy internal coating
- Insulation : none
- Heating : internal serpentine
- Mixing : gas recirculation (gas blower)

COST : -

FEEDSTOCK : Industrial wastes of wheat starch factory

TREATMENT : None

RUNNING CONDITIONS :

B_V = 3.6 kg COD m^{-3} d^{-1}

T = 35-37 °C

θ_H = 4.6 d

θ_S = 4.6 d

GAS :
- Storage : fixed tank HGAD space (mild steel)
- Production : rate : 1.11 m^3 d^{-1} m^{-3}
  : yield : 0.31 m^3 kg^{-1} COD_o
- Composition : -
- Depollution : -
- Uses : 100 % hot water (professional use)

EFFLUENT :
- Treatment : ultrafiltration
- Uses : discharge to sewer

MAJOR PROBLEMS :
None for the moment

D-BRAUNSCHWEIG - DIGESTER 87

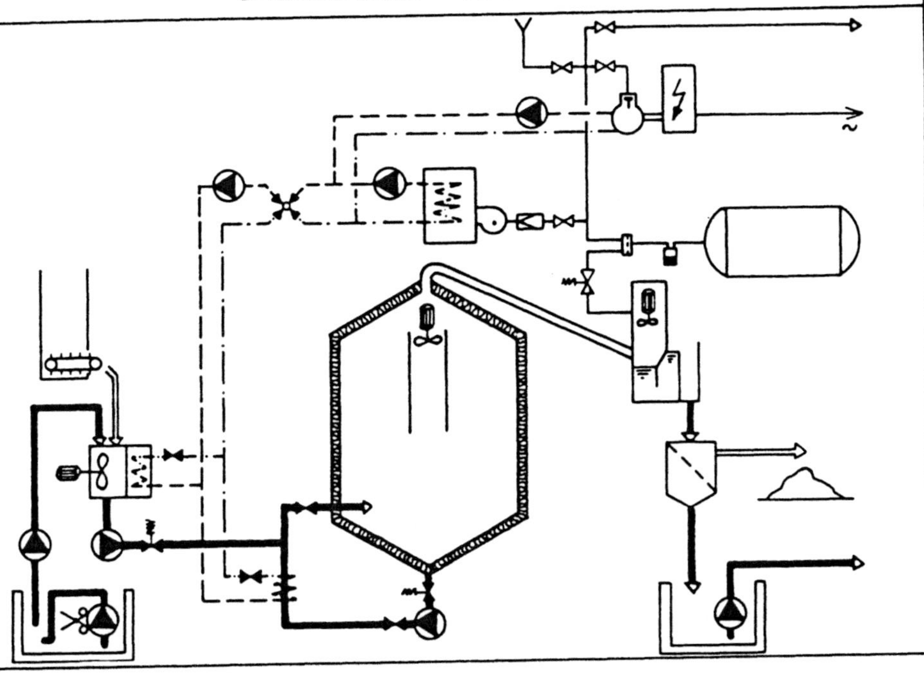

ADDRESS : Institut für Technologie, Bundesforschungsanstalt für Landwirtschaft (FAL), Bundesallee 50,
D-3300 BRAUNSCHWEIG, Federal Republic of Germany

START-UP : 07.1980

DESIGN OF THE DIGESTER :

- System : partially mixed, intermittent loading
- Volume : total = 100 m^3, working volume = 100 m^3
- Material : steel
- Insulation : polyurethane foam
- Heating : 2 external heat exchangers
- Mixing : internal propeller installed in the central tube, acting downwards + centrifugal pump

COST : 1 200 000 DM (total)

FEEDSTOCK : Liquid manure from cows, heifers, calves (contin.), straw and vegetable material (freq.)

TREATMENT : Liquid manure stored for max. 3 days, straw chopped, veg. material fresh or ensiled

RUNNING CONDITIONS :

B_V = 2-5 kg VS m^{-3} d^{-1}

T = 30-50 °C

θ_H = 10-25 d

θ_S = 10-25 d

GAS :

- Storage : gas holder diaphragm type (75 m^3)
- Production : rate : 0.8-2.0 m^3 d^{-1} m^{-3}
  : yield : 0.38-0.42 m^3 kg^{-1} VS_o
- Composition : 59-62 % CH_4
- Depollution : 35-40 % VS reduction
- Uses : 100 % for electricity production

EFFLUENT :

- Treatment : separation and heaping up coarse solids, liquid stored up to 8 weeks
- Uses : solids as soil conditioner, liquid as fertilizer

CH-UNTERENDINGEN - DIGESTER 136

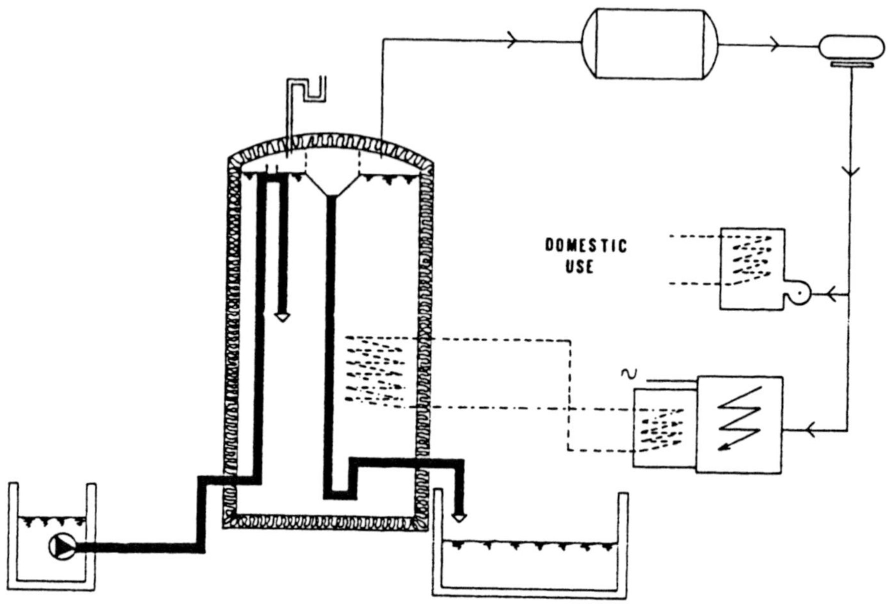

ADDRESS : K. Meier, Steig, 4
CH-5305 UNTERENDINGEN, Switzerland

START-UP : 1978

DESIGN OF THE DIGESTER :
- System : partially mixed, intermittent loading
- Volume : working volume = 66 m^3
- Material : glass fibre reinforced plastic, int. coating : neoprene
- Insulation : polymeric (10-12 cm)
- Heating : internal heat exchanger
- Mixing : none

COST : Unknown

FEEDSTOCK : Manure from 40 beef cattle + 9 dairy cattle the whole year, fed with raw fodder (dairy) and silage (beef) in summer and silage in winter for both

TREATMENT : None

RUNNING CONDITIONS :
B_V = 1.1 kg VS m^{-3} d^{-1}
T = 33 °C
θ_H = 51 d
θ_S = -

GAS :
- Storage : balloon in polymeric material
- Production : rate : 0.62 m^3 d^{-1} m^{-3}
  : yield : 0.56 m^3 kg^{-1} VS_o
- Composition : 59 % CH_4
- Depollution : 40 % VS reduction

Uses : internal use + domestic use (hot water + heating + electric power)

EFFLUENT :
- Treatment : none
- Uses : soil conditioner

MAJOR PROBLEMS :
None for the moment

B-BAILLONVILLE - DIGESTER 281

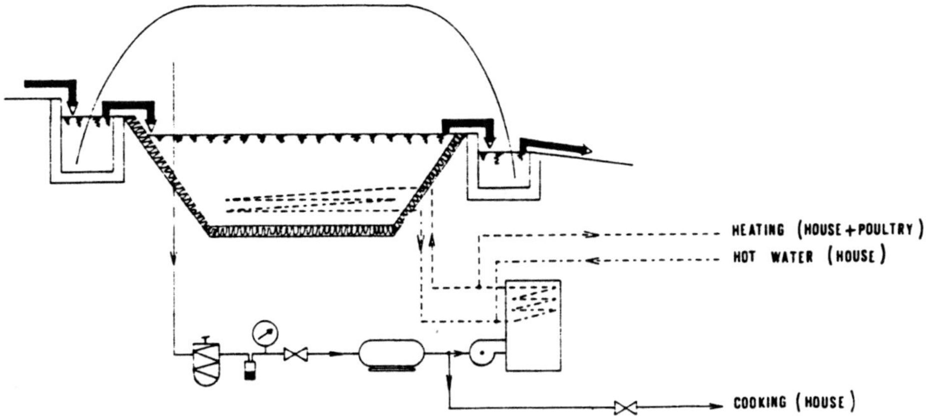

ADDRESS : M. Lhermite, chemin du Tombeu, 4
B-5413 BAILLONVILLE, Belgium

START-UP : 11.1980

DESIGN OF THE DIGESTER :

- System : plug-flow, intermittent loading, one step
- Volume : total = 45 m^3, working volume : 21 m^3
- Material : polymeric (polystyrene)
- Insulation : polystyrene plates + mineral wool (18 cm)
- Heating : internal serpentine
- Mixing : none

COST : 585 000 BFR

FEEDSTOCK : Manure from 850 chickens (+ sometimes antibiotics)

TREATMENT : Storage of 1 day + dilution

RUNNING CONDITIONS :

B_V = 1.8 kg VS m^{-3} d^{-1}
T = 35 °C
θ_H = 25 d
θ_S = 25 d

GAS :

- Storage : compressor + balloon inside the digester.
- Production : rate : 0.86 m^3 d^{-1} m^{-3}
  : yield : 0.48 m^3 kg^{-1} VS_o
- Composition : 56 % CH_4
- Depollution : -
- Uses : 30 % internal use + 70 % hot water (domestic use) + heating (domestic + professional use) + cooking (domestic use)

EFFLUENT :

- Treatment : none
- Uses : soil conditioner

MAJOR PROBLEM :

Clogging of outlet

DK-HASLEV - DIGESTER 6

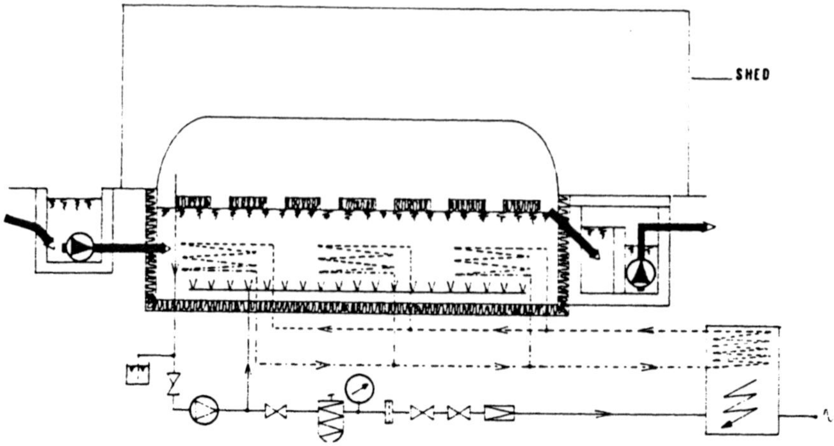

ADDRESS : Assendrup Hovedgaard, Assendrupvej, 10, Aversi
DK-4690 HASLEV, Denmark

START-UP : 08.1979

DESIGN OF THE DIGESTER :

- System : plug-flow, intermittent loading
- Volume : total = 2 x 220 m^3, working volume = 2 x 200 m^3 (in parallel)
- Material : concrete
- Insulation : Leca (granular expanded clay) + polymeric (top)
- Heating : internal serpentine
- Mixing : gas recirculation (installed : 10.1981)

COST : ± 800 000 DKR excl. VAT

FEEDSTOCK : Manure from 165 cows + 15 heifers + 30 calves (1-2 years) + 80 calves (≤ 1 year) + 84 calves for slaughtering (= 210 livestock units) fed with compounds and silage the whole year

TREATMENT : Storage only in the channels of the stables

RUNNING CONDITIONS :

B_V = 2.9 kg VS m^{-3} d^{-1}
T = 33-35 °C
θ_H = 25 d
θ_S = 25 d

GAS :

- Storage : balloons (2 x 150 m^3) in papyrotex 1 mm
- Production : rate : 0.63 m^3 d^{-1} m^{-3}
  : yield : 0.22 m^3 kg^{-1} VS_o
- Composition : 65 % CH_4
- Depollution : -
- Uses : 100 % for electricity production (19.6 % as electricity + 65 % hot water for heating of slurry)

EFFLUENT :

- Treatment : storage up to 6 weeks
- Uses : soil conditioner applied by irrigation

MAJOR PROBLEMS :

Loss in organic content during storage, scum formation, inefficient mixing device, short-circuiting of feed-flow, electricity generation engine problems. All these problems solved during 1982, improvement in gas production from 156 m^3/day to 250 m^3/day

F-POUCHARRAMET - DIGESTER 296

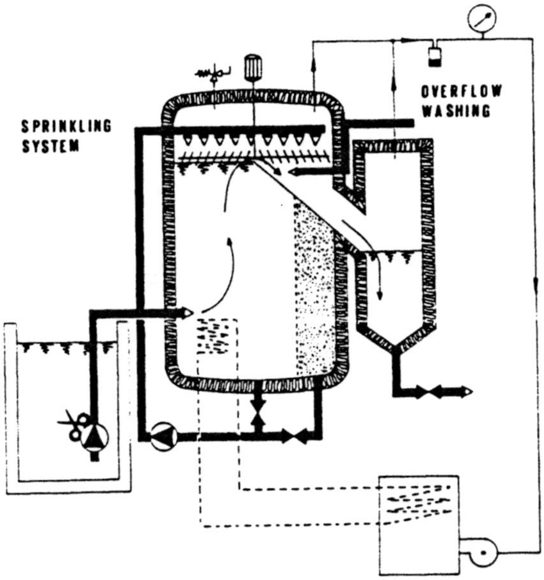

ADDRESS : Domaine de Borret
F-31370 POUCHARRAMET, France

START-UP : 06.1981

DESIGN OF THE DIGESTER :

- System : plug-flow on solid matter, intermittent loading, pilot scale
- Volume : total = 5 m^3, working volume = 3.8 m^3
- Material : mild steel, internal coating : polyurethane paint
- Insulation : aluminium covered glass wool (15 cm)
- Heating : internal serpentine + external electric heater
- Mixing : liquid recirculation (pump)

COST : Experimental plant

FEEDSTOCK : Manure from dairy cattle fed with silage the whole year (+ Cu + antibiotics)

TREATMENT : Dilution + grinding

RUNNING CONDITIONS :

B_V = 6.7 kg VS m^{-3} d^{-1}

T = 35 $\pm$ 1 °C

θ_H = 17 d

θ_S = -

GAS :

- Storage : none
- Production : rate : 1.8 m^3 d^{-1} m^{-3}
  : yield : 0.275 m^3 kg^{-1} VS_o
- Composition : 57 % CH_4
- Depollution : 40 % VS reduction
- Uses : internal use (theorical values) :
  34 % in winter, 18 % in summer

EFFLUENT :

- Treatment : none
- Uses : soil conditioner

MAJOR PROBLEMS :

None for the moment

F-RENESCURE - DIGESTER 534

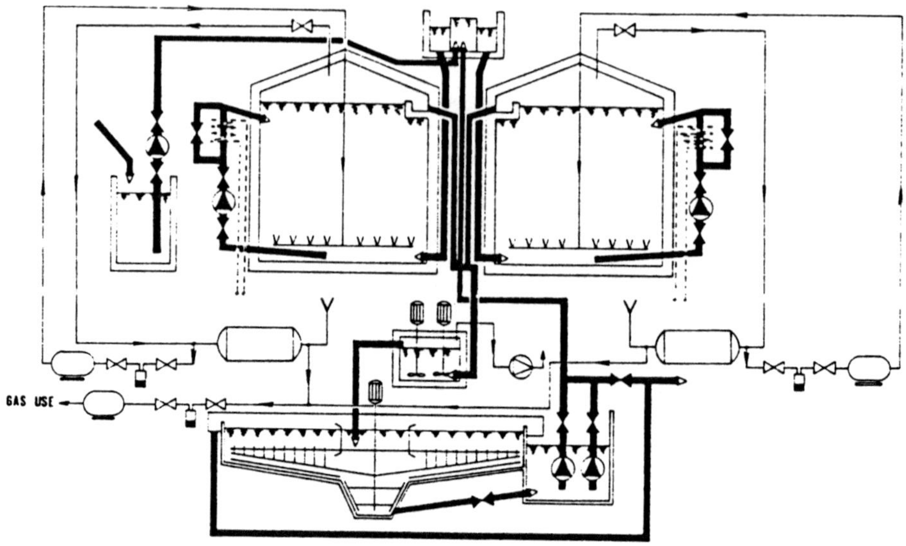

ADDRESS : Ets. Bonduelle, usine de Renescure
F-59173 RENESCURE, France

START-UP : 07.1981 + 04.1982

DESIGN OF THE DIGESTER :

- System : continuous, anaerobic contact
- Volume : total = 2 x 2 800 m^3, working volume = 2 x 2 500 m^3
- Material : steel, int. coating : black paint, ext. coating : grey paint
- Insulation : double envelope of Alu. with 5-6 cm air for 1 digester
- Heating : external heat exchanger (in fact, cooling system)
- Mixing : gas recirculation, compressor

COST : 7 000 000 FF (1980)

FEEDSTOCK : Vegetable canning industry waste-waters

TREATMENT : None : separative channels in the plant

RUNNING CONDITIONS :

$B_V = \pm 2$ kg COD m^{-3} d^{-1}
4.2 kg COD m^{-3} d^{-1}
(= maximum)

T = 35 $\pm$ 5 °C

θ_H = 5 d

θ_S = 15 d

GAS :

- Storage : balloon in polymeric material
- Production : rate : $\pm$ 1.6 m^3 d^{-1} m^{-3}(1.8 max.)
  : yield : 0.43-0.50 m^3 kg^{-1} COD_o
- Composition : 60 % CH_4
- Depollution : $\pm$ 90-95 % COD reduction
- Uses : 100 % steam 10 bar for professional use

EFFLUENT :

- Treatment : anaerobic lagooning (one part of the year)
- Uses : fertilizer (other part of the year)

MAJOR PROBLEMS :

None for the moment

F-LECOUDRAY MONTCEAU - DIGESTER 487

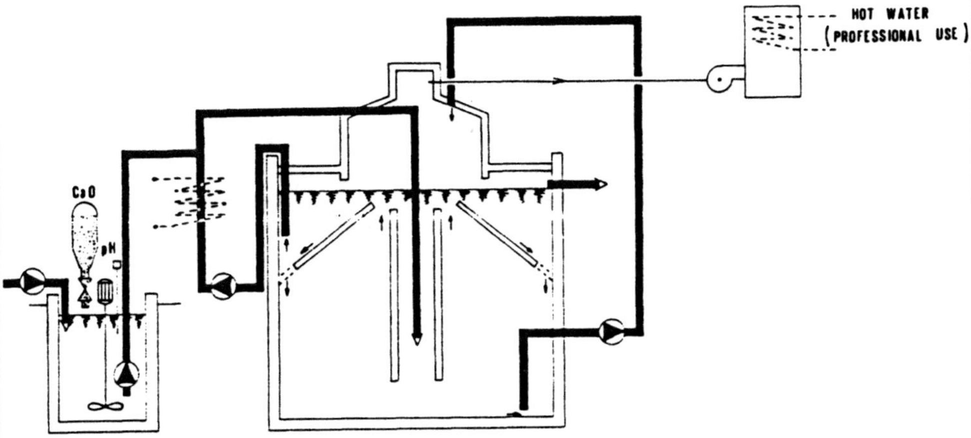

ADDRESS : Distillerie Agricole de la Région d'Auvernaux
F-91830 LECOUDRAY MONTCEAU, France

START-UP : Autumn 1981

DESIGN OF THE DIGESTER :

- System : continuous, anaerobic contact in one tank (IRIS)
- Volume : total = 1 500 m^3, working volume : 1 200 m^3
- Material : concrete (mainly)
- Insulation : none
- Heating : external heat exchanger
- Mixing : liquid recirculation (pump) and mechanical (agitation axle)

COST :1 500 000 FF (1981)

FEEDSTOCK : Distillery slops wastewater

TREATMENT : Low dilution + lime neutralization

RUNNING CONDITIONS :

$B_V = \pm 4.2$ kg COD m^{-3} d^{-1}

T = 35 °C

$\theta_H = \pm 3$ d

θ_S = ?

GAS :

- Storage : none
- Production : rate : $\pm$ 1.3 m^3 d^{-1} m^{-3}
  : yield : 0.30 m^3 kg^{-1} COD_o
- Composition : $\pm$ 60 % CH_4
- Depollution : $\pm$ 80 % COD reduction
- Uses : 100 % hot water for professional use

EFFLUENT :

- Treatment : flocculation
- Uses : soil conditioner

MAJOR PROBLEMS :

Problems with internal mixing

IRL-GALWAY - DIGESTER 236

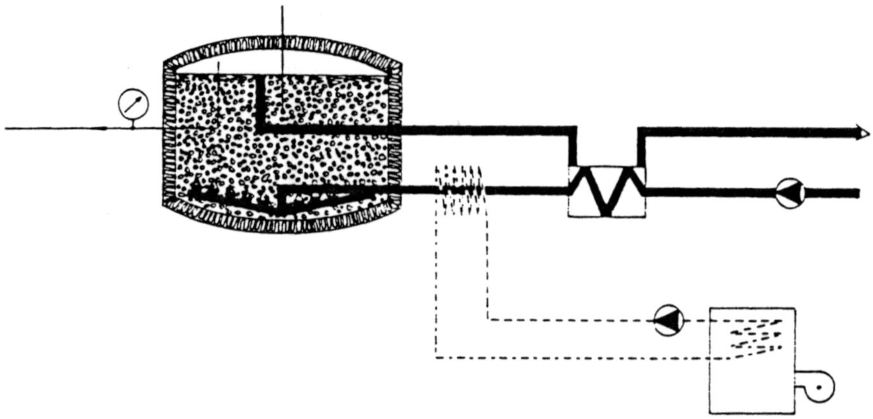

ADDRESS : Agricultural College, Athenry
IRL - Co. GALWAY - Ireland

START-UP : 27.04.1982

DESIGN OF THE DIGESTER :

- System : anaerobic filter + plastic matrix (poros. 0.94), pilot scale
- Volume : total = 3.5 m^3, working volume = 2.64
- Material : fibreglass
- Insulation : expanded polyurethane (10 cm)
- Heating : external heat exchanger
- Mixing : none

COST : IRL 3 000 approx.

FEEDSTOCK : Manure from 50 sows and 50 weaners fed with compounds the whole year (+ antibiotics)

TREATMENT : Dilution by rainwater

RUNNING CONDITIONS :

B_V : 3.7-4.7 kg COD m^{-3} d^{-1}

T = 25 °C

θ_H : 5.5 d

θ_S = ?

GAS :
- Storage : none (vented to the atmosphere)
- Production : rate : 0.95-1.21 m^3 d^{-1} m^{-3}
  : yield : 0.26 m^3 kg^{-1} COD_o
- Composition : 86 % CH_4
- Depollution : 66 % COD reduction
- Uses : none

EFFLUENT :
- Treatment : none
- Uses : none

MAJOR PROBLEMS :
None for the moment

I-ARGELATO - DIGESTER 472

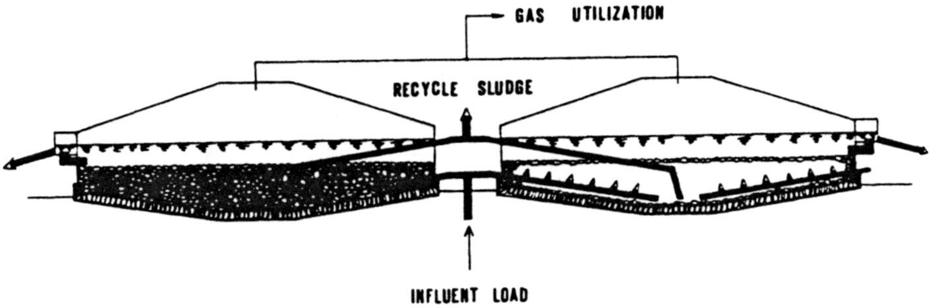

<u>ADDRESS</u> : Societa Italiana Industria Zuccheri
SIIZ - via Costituzione, 3
I-FUNO di ARGELATO (Bologna), Italy

START-UP : -

DESIGN OF THE DIGESTER :
- System : double digester, upflow anaerobic filter, continuous
- Volume : total = 2 x 1 638 m^3, working volume = 2 x 1 248 m^3
- Material : steel
- Insulation : polymeric material
- Heating : external heat exchanger
- Mixing : none

COST : -

FEEDSTOCK : Wastewater from industrial sugar plant (12 000 t d^{-1} sugar beets)

TREATMENT : COD reduced from 22-45 g/l to 17 g/l and pH brought from 4-12 to neutral

RUNNING CONDITIONS :

B_V = 2 x 8.2 kg COD m^{-3} d^{-1}

T = 35 $\pm$ 1 °C

Θ_H = 1.17 d

Θ_S = 1.17 d

GAS :
- Storage : floating gasholder (250 m^3)
- Production : rate : 2 m^3 d^{-1} m^{-3}
  : yield : 0.245 m^3 kg^{-1} COD
- Composition : 70 % CH_4
- Depollution : 70 % COD reduction
- Uses : digester heating and process

EFFLUENT :
- Treatment : aerated lagoon
- Uses : discharge

MAJOR PROBLEMS :
Not available

CH-LINDAU - DIGESTERS 131-132

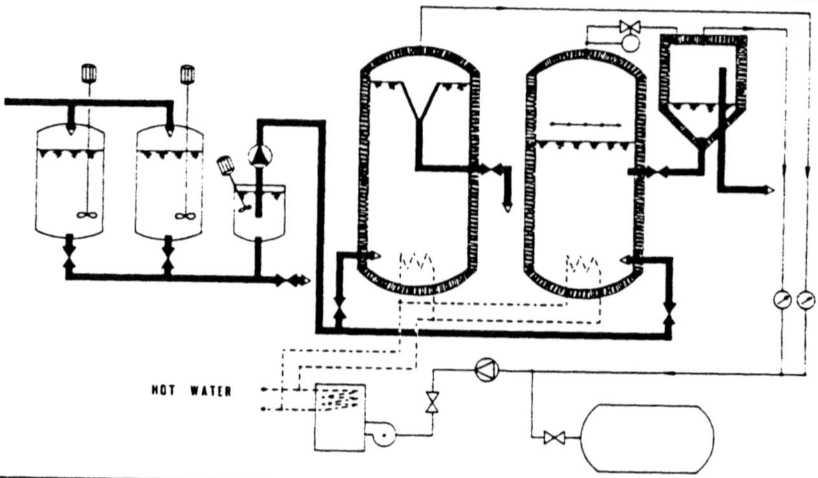

ADDRESS : Landwirtschaftliche Schule Strickhof
CH-8307 LINDAU, Switzerland

START-UP : -

DESIGN OF THE DIGESTER :

- System : dig. 1 : upflow, not mixed, intermittent loading
  dig. 2 : mixed by evolving gas, scum retention, int. loading
- Volume : dig. 1 : 6.85 m^3
  dig. 2 : 8.16 m^3 (working volume)
- Material : steel cylinders
- Insulation : polyurethane foam (outside) : 15 cm
- Heating : internal serpentine
- Mixing : dig. 1 : some mixing by thermal convection
  dig. 2 : digestion liquor is forced into an expansion vessel by the pressure of the gas. At a predetermined pressure, gas valve opens which produces mixing by back-flow of the digestion liquor

COST : -

FEEDSTOCK : Bovine cattle manure (4.7 % TS; 3.5 % VS)

TREATMENT : Depends on experiments with each digester : tank of 10 d feed capacity, where special substrate mixtures are prepared

RUNNING CONDITIONS :

B_V = 2.1 kg VS m^{-3} d^{-1}
T = 34 $\pm$ 0.5 °C
θ_H = 16 d
θ_S = ? (tracer studies going on)

GAS :

- Storage : synthetic rubber balloon (15 m^3)
- Production : rate : (m^3 m^{-3} d^{-1})
  1) 0.75; 2) 0.86
  : yield : (m^3 kg^{-1} VS_o)
  1) 0.34; 2) 0.39
- Composition : 62 % CH_4
- Depollution : -
- Uses : digester heating; warm water

EFFLUENT :

- Treatment : not available
- Uses : not available

MAJOR PROBLEMS : Not available

NL-KOOG a/d ZAAN - DIGESTER 506

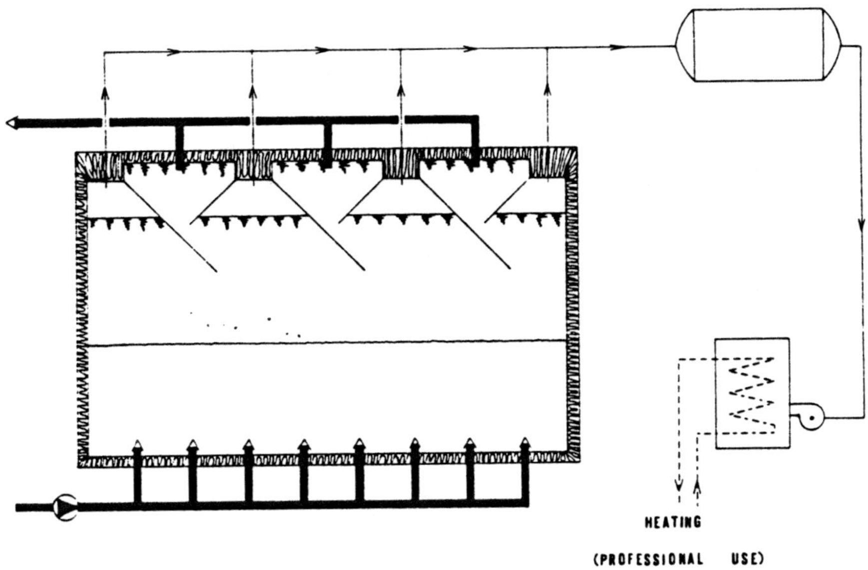

ADDRESS : Z.B.B., Zetmeelbedrijven
NL-KOOG a/d ZAAN, The Netherlands
START-UP : 1980
DESIGN OF THE DIGESTER :
- System : upflow without carrier, continuous loading
- Volume : total = 800 m^3
- Material : concrete
- Insulation : polymeric
- Heating : none
- Mixing : none

COST : 2 000 000 HFL

FEEDSTOCK : Maize starch factory wastewater
TREATMENT : None
RUNNING CONDITIONS :
B_V = 11 kg COD m^{-3} d^{-1}
T = 35 °C
θ_H = 0.65 d
θ_S = -

GAS :
- Storage : gasholder in galvanized iron
- Production : rate : 3.0 m^3 d^{-1} m^{-3}
  : yield : 0.27 m^3 kg^{-1} COD_o
- Composition : 70 % CH_4
- Depollution : 85 % COD reduction
- Uses : 100 % heating (professional use)

EFFLUENT :
- Treatment : freshing up by cascade
- Uses : discharge

MAJOR PROBLEMS :
- Precipitation in pipelines as a result of high alkalinity in the water
- Sulphite toxicity

I-CAROBBIO - DIGESTER 204

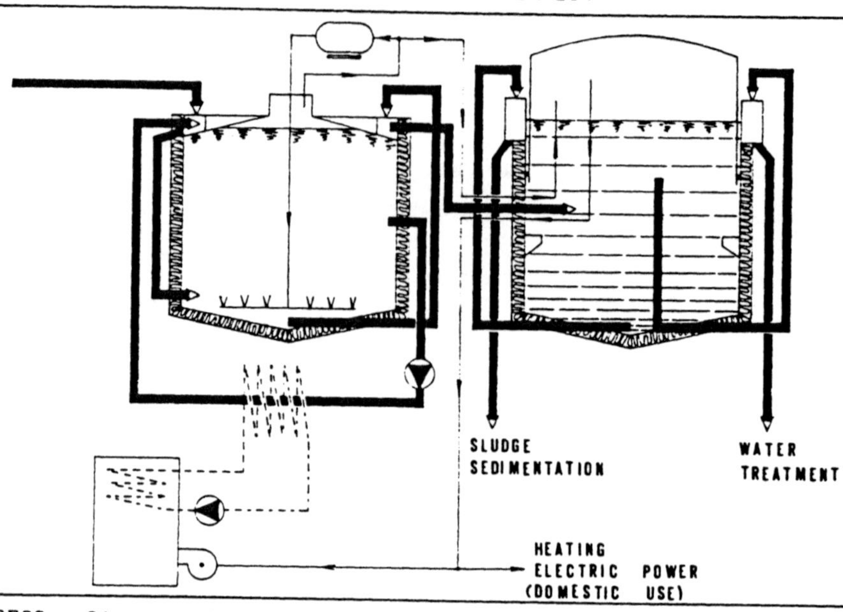

ADDRESS : Giuseppe Citterio S.p.a.
I-CAROBBIO D.A. (BG) - Italy

START-UP : 12. 1980

DESIGN OF THE DIGESTER :

- System : two-step, completely mixed digester, intermittent loading
- Volume : total = 900 + 450 m^3, working volume = 800 + 400 m^3 (in series)
- Material : concrete
- Insulation : mineral wool (5 cm)
- Heating : external heat exchanger
- Mixing : gas recirculation (pump)

COST : 300 000 000 LIT

FEEDSTOCK : Manure from 2500 swines fed with compounds the whole year (+ Cu + antibiotics)

TREATMENT : Grinding

RUNNING CONDITIONS :

B_V = 1.5 kg VS m^{-3} d^{-1}

T = 35 °C

θ_H = 20 + 10 d

θ_S = -

GAS :

- Storage : floating gasholder in mild steel
- Production : rate : 0.67 m^3 d^{-1} m^{-3}
  : yield : 0.43 m^3 kg^{-1} VS_o
- Composition : 67 % CH_4
- Depollution : 66 % VS reduction
- Uses : 25 % internal use + 75 % domestic use (heating + electric power)

EFFLUENT :

- Treatment : decantation (compost with straw)
- Uses : soil conditioner

MAJOR PROBLEMS :

Groundwater or rainwater entry and settlement during storage, problems with gas metering devices

UK-STEWARTBY - DIGESTER 456

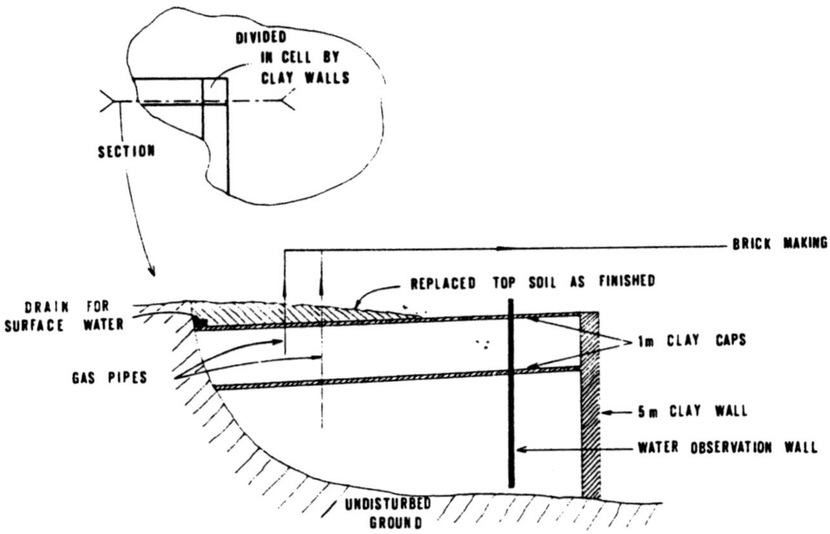

<u>ADDRESS</u> : London Brick Landfill Ltd.
UK-STEWARTBY (Bedford), MK439LZ - United Kingdom

<u>START-UP</u> : 1979

<u>DESIGN OF THE DIGESTER</u> :
- System : landfill
- Volume : 72 ha x 25 m depth, divided in cells of 2.5 ha
- Material : clay
- Insulation : none
- Heating : self-generating
- Mixing : none

<u>COST</u> : UKL 250 000

<u>FEEDSTOCK</u> : Raw refuse

<u>TREATMENT</u> : Special treatment for difficult toxic wastes

<u>RUNNING CONDITIONS</u> :
B_V = 1 100 ton year^{-1}
T = 20-30 °C
θ_H = permanent
θ_S = -

<u>GAS</u> :
- Storage : none
- Production : rate : up to 5 300 m^3 d^{-1}
  : yield : 400 m^3 ton^{-1} (hoped for)
- Composition : 60 % CH_4
- Depollution : -
- Uses : 100 % heating (brick making)

<u>EFFLUENT</u> :
- Treatment : pump off leachate + will be treated aerobically in time
- Uses : discharge to land

<u>MAJOR PROBLEMS</u> :
None for the moment

CHAPTER II. HARDWARE : THE ENGINEERING ASPECTS OF BIOGAS PLANTS

The introductory chapter defined a biogas plant and more particularly a methane digester. A methane digester is a closed vessel, gas-tight at least at its top, insulated in most cases, and equipped with a heating and a mixing device. Different biomethanation systems were described, systems which were essentially a function of the types of feedstocks that are treated. In this chapter on hardware, these aspects are discussed in greater detail.

II.1. Overview of the methane digestion systems

II.1.1. Types of methane digestion systems appropriate to particular wastes

Table I.2.1. Methane digestion system used as a function of the type and nature of treated waste

Methane digestion system	Share among wastes (as %)		
	Liquid (1) agricultural wastes	Solid (2) agricultural wastes	Industrial (3) wastewaters
Completely mixed	68	16	14
Partially mixed	4		
Plug-flow	16	27	
Batch	4	49	
Fed-batch	3		
Two-step	1	4	2
Accumulation	1		
Anaerobic filter	2	4	15
Anaerobic contact	1		19
IRIS system			9
Upflow without carrier			41
	100	100	100

(1) 334 methane digestion systems identified among 358 methane digesters on said waste; (2) 55 methane digester systems identified among 56 methane digesters on said waste; (3) 80 methane digester systems identified among 84 methane digesters on said waste.

The methane digestion systems are essentially a function of the type and nature of the waste to be treated. This relation is detailed in Table I.2.1 for liquid and solid agricultural wastes and for industrial wastewaters.

The completely mixed methane digestion system is obviously the most widely used for the more liquid agricultural wastes, independently of the type of manure considered : bovine, cattle, pig, poultry, etc. The plug-flow methane digestion system is also in general use but as yet to a lesser extent. The oldest systems, namely the batch and fed-batch systems are still in use for liquid or semi-solid manures. In Switzerland there is even a revival of interest in the fed-batch system. This system can easily be applied, simply by insulation and minor transformation of existing manure storage tanks. Although these types of methane digesters have a low biogas production, since they operate at psychrophilic temperatures, they present the advantage of very low investment cost.

The more advanced technologies such as the anaerobic filter, the anaerobic contact, or other systems with accumulation of biomass inside the methane digester have been little used and when they are used, it is generally for the methane digestion of pig manure at the pilot-scale level.

As regards the biomethanation systems for solid agricultural wastes, the batch system is still preferred. The completely mixed and the plug-flow systems are also used but, in these cases, the solid waste must be diluted and/or ground before biomethanation.

As far as the biomethanation of industrial wastewaters is concerned, more advanced technologies based on recycling or trapping of active biomass are in general use. Forty-one per cent of the wastewaters are treated by the upflow system without carrier, well known as the UASB process. Most of these UASB systems are located in The Netherlands where the system originated (Lettinga *et al*., 1980). Nineteen per cent of the industrial biogas plants are based on the anaerobic contact process. The IRIS process which can be assimilated to the anaerobic contact process concerns 9 % of the biomethanation systems on industrial wastewaters. These IRIS methane digesters are concentrated in the north of France and in Belgium, close to their place of origin (Devillers *et al*., 1977). The completely mixed system is still in use, especially in Italy for the biomethanation of distillery wastewaters. Finally, the anaerobic filter which accounts for 14 % of the industrial biogas plants is until now essentially in use at the pilot-scale level, although the first full-scale plants are reported to be on stream in Italy and France.

II.1.2. Engineering aspects (design and construction of biogas plants)

a. Materials used for the construction of methane digesters

The three major categories of materials used for the methane digestion tank are the following :

- concrete or assimilated *i.e.* reinforced concrete or concrete with some parts in polymeric material or steel,

- steel,
- polymeric material or assimilated, i.e. glassfibre-reinforced plastic.

As shown in Fig. I.2.1, steel and concrete are the most widely used, independently of the size of the methane digester. With the exception of digester sizes below 50 m^3 of digester working volume, polymeric materials are little utilized and they are never used for sizes larger than 1 000 m^3 of digestion working volume.

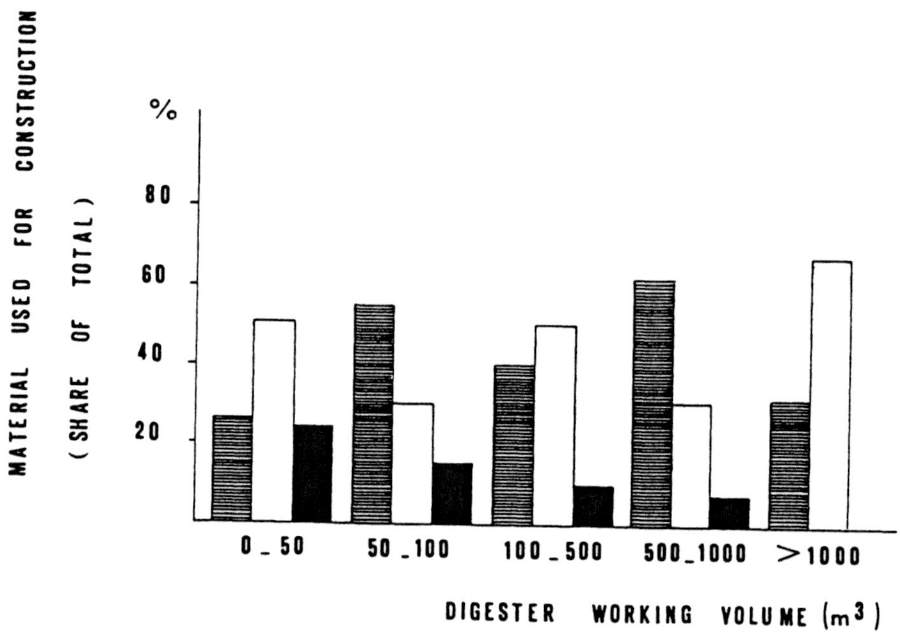

Fig. I.2.1. Type of material used for the construction of a methane digester as a function of its size

▤ : Concrete; ☐ : Steel; ■ : Polymeric material.

b. Materials employed for the insulation of methane digesters

As will be seen later, most of the methane digesters, both agricultural and industrial, are operating under mesophilic conditions. Therefore methane digesters need to be insulated in order to minimize heat losses.

Two important types of insulation material may be distinguished. The first type is of synthetic origin and comprises all polymeric materials such as polyurethane or polystyrene. The second type is of natural origin and comprises glass wool, mineral or rock wool, earth, granular expanded clay and vermiculite (a variety of mica). The frequency of use of these different types of insulation materials encountered in the methane digesters surveyed is reported in Table I.2.2.

Table I.2.2. Frequency of use of insulation materials

	Share of total (%)
1° Insulation material of synthetic origin	
- Polymeric material (no specification of the type)	41
- Expanded polyurethane	9
- Expanded or extruded polystyrene	14
2° Insulation material of natural origin	
- Glass wool	10
- Mineral or rock wool	13
- Earth	2
- Granular expanded clay	1
- Vermiculite (variety of mica)	1
3° No insulation	9

From Table I.2.2, it appears that insulation materials of synthetic origin are preferred to those of natural origin. The reason for this is that the former are easier to apply and moreover they generally have a better specific insulation capacity.

It should be noted that all these insulation materials are coated with a waterproof cladding such as aluminium sheets. Indeed, any ingress of moisture must be avoided since it reduces the insulation properties. Problems of this order were reported in Switzerland.

In Table I.2.2, it can be seen that there is no insulation at all in some cases. This is the case for some agricultural methane digesters operating under psychrophilic conditions. Since these digesters are transformed storage tanks, they cannot be insulated easily. Some digesters, operating only during the summer, also present this particularity. Some industrial methane digesters also are not insulated, the wastewaters that are to be treated being so hot that neither insulation nor heating device are necessary.

c. Heating devices of methane digesters

In order to heat the influent and to maintain the digester content at the required temperature, various types of heating devices are in use. The most currently encountered heating devices are the following :

- a heat exchanger which heats the digester content, directly when it is located inside the digester, by recirculation when it is located outside the digester. This device is more convenient for wastes with a very low solid content;

- heating coils or heating draft tubes located inside the digester, in which warm water is circulated to heat the digester content. This device is applicable to any type of waste;
- jacketed vessels which use hot water surrounding the digestion tank. This device is generally applied to pilot-scale methane digesters.

These three different types of heating devices are shown schematically in Fig. I.2.2.

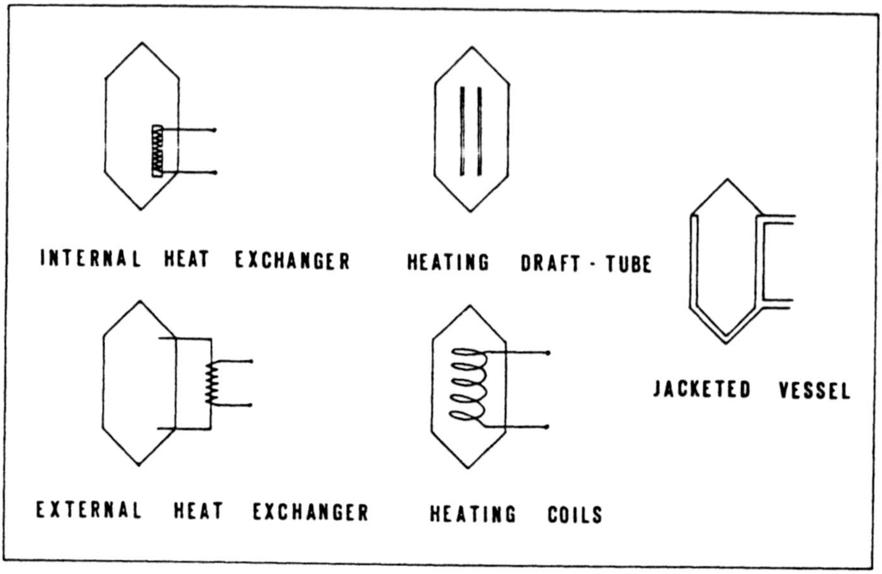

Fig. I.2.2. Heating devices of methane digesters

The percentage distribution of the various heating devices utilized in the surveyed methane digesters is presented in Table I.2.3. It will be observed that heating coils and heat exchangers represent the majority of the heating devices. It appears also that internal heating devices are preferred to external ones. This is obviously aimed at minimizing heat losses, but leads to problems when repairs are needed.

Alternative heating devices are also used under particular circumstances. Electric heaters are encountered in "do it yourself" methane digesters where existing used material has been re-utilized for the construction of the digester. Heating by steam injection is applied in industrial methane digesters where waste steam is available. Finally, heating by placing the digester inside a heated room or a greenhouse is applied only for very small digesters such as pilot-scale ones.

Table I.2.3. Heating devices of methane digesters

	Share of total (%)
- Internal heating devices	
. Heating coils	32
. Heat exchanger	27
. Electric heater	3
. Steam injection	2
	64
- External heating devices	
. Heat exchanger	25
. Jacketed vessel	7
. Heating coils	2
. Heated room or greenhouse	1
	35
- Internal and external heating devices	1

d. Mixing devices of methane digesters

In the introductory chapter, it was seen that some biomethanation systems such as the completely mixed, the anaerobic contact and the IRIS systems appeared to require mixing. Experience has shown that mixing is essential to avoid scum formation and settlement in the reactor.

The three important mixing devices : liquid recirculation, gas recirculation, and mechanical agitation, are schematized in Fig. I.2.3.

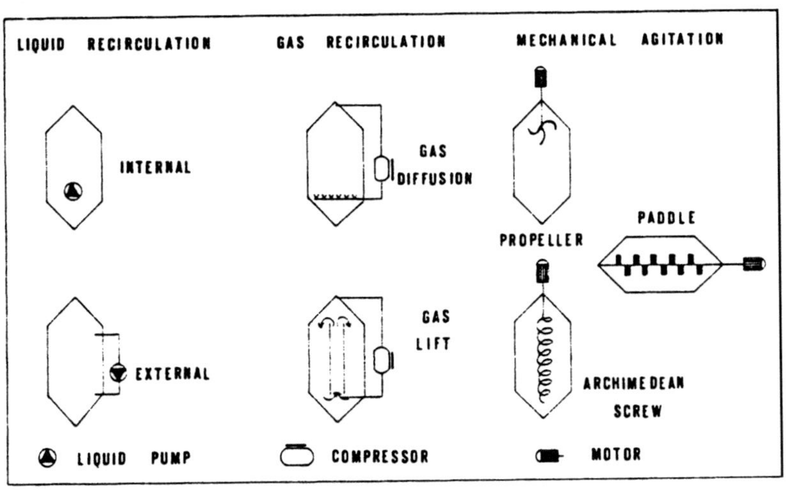

Fig. I.2.3. Mixing devices of methane digesters

The liquid recirculation system can be either internal or external, internal when an immersed pump mixes the digester content and external when the digester content is recirculated from the upper part to the bottom of the digester or vice versa. Mixing by gas recirculation is achieved by the injection of compressed gas either at the bottom of the digester, or through a vertical draft tube of relatively large diameter. This latter is called a "Gas-Lift" system and is often used in sewage digesters and in industrial digesters. Finally, the mechanical agitation system can be applied by using paddles, propellers or archimedean screws. Paddles are often used in plug-flow digesters whereas propellers were encountered as an additional mixing device in Danish digesters.

Table I.2.4. Mixing devices of methane digesters

	Share of total (%)
- Liquid recirculation	35
- Gas recirculation	25
- Mechanical mixing	
. Paddle	12
. Propeller	5
. Archimedean screw	2
- Liquid and gas recirculation	4
- Liquid recirculation and mechanical mixing	14
- Gas recirculation and mechanical mixing	3

These different mixing devices have been encountered in the digesters surveyed, either alone or combined. Their frequency of use is reported in Table I.2.4. Mixing by liquid or gas recirculation is the most frequently practised, independently of the scale of the methane digester. On the other hand, mixing by mechanical agitation is essentially encountered in small digesters with a digestion working volume less than or equal to 100 m^3.

e. Gas storage systems

When produced, the biogas needs to be stored before its further utilization. Two gas storage systems are possible : the first at variable pressure and fixed volume, the second at variable volume and fixed pressure. The latter is most frequently encountered in practice. In general, a preference is observed for the plastic membrane type gasholders, known as gas balloons, and for floating gasholders. These two types of gasholders are sometimes placed inside the digester. High pressure tanks were the gas storage systems least encountered during the survey.

Gas balloons made of polymeric material are convenient for utilization and are not an expensive investment. Nevertheless, they need to be protected against wind and rodents. The protective materials commonly employed are rubber cloths or iron sheets. Floating gasholders are generally made of steel and are more durable, although precautions must be taken against freezing of the water seal.

Megara (Greece) 1983. Water-seal gasholder of a 500 m^3 anaerobic contact digester.

Thorembais-les-Béguines (Belgium) 1982. 50 m^3 baloon gasholder.

f. Biogas purification systems

Prior to its use, the biogas may be scrubbed. Gas dewatering is usually applied, but the removal of the more acid gases present as impurities in the biogas is less frequent.

Water is probably the most undesirable impurity in biogas streams. In addition to enhancing corrosion problems, water vapour can condense whenever the biogas is cooled or compressed. Moreover, the presence of high levels of moisture also reduces the heating value of the gas and adds to the transmission load of the pipelines.

Water removal can be accomplished in many ways : condensation by compression and/or cooling, adsorption on activated solid dessicants or absorption by hygroscopic liquids. Table I.2.5 recapitulates the different dewatering systems in use in the surveyed countries.

Table I.2.5. Biogas scrubbing systems

1° Dewatering
- Condensation traps or pots (at ambient T) - Passage on dessicants (salts) - Filtration on ceramic rings - Gravel filtration (grit filtration) - Frigorific dewatering
2° Hydrogen sulphide scrubbing
- Water washing/trapping - Passage on iron filings or passage on iron oxide/wood shavings - Addition of $FeCl_3$ to manure - Filtration : Landre-Ruhaak system
3° Carbon dioxide scrubbing
- Water trapping - Drawers of quicklime - Catalysol

The acid gases, carbon dioxide and hydrogen sulphide, may have to be removed for corrosion prevention. Moreover, when the levels of hydrogen sulphide are high, the scrubbing of hydrogen sulphide becomes obligatory to meet the safety regulations on permitted levels.

The hydrogen sulphide removal processes may be classified in two categories : dry oxidation processes, and liquid phase oxidation processes. Such processes encountered in the survey are reported in Table I.2.5. Finally, although carbon dioxide is seldom scrubbed, three different systems were elicited by the survey : water trapping where the carbon dioxide is eliminated as carbonic acid; drawers filled with quicklime where the carbon dioxide is transformed into calcium bicarbonate; and catalysol.

CHAPTER III. SOFTWARE : BIOTECHNOLOGICAL ASPECTS

This chapter considers the performances, i.e. the biogas productivities (rates and yields) obtained in the methane digesters surveyed. As will be seen, these biogas productivities vary not only in function of the type of waste considered and the type of biomethanation system used but also in function of the running conditions. The first section of this chapter is therefore devoted to the analysis of running conditions whereas in the second section the biogas productivities are discussed in detail.

III.1. Running conditions

Besides the fact that different biomethanation systems can be used for the treatment of the various feedstocks, the running conditions under which biomethanation occurs can also vary.

Three different running parameters influence the biogas productivities : the volumetric load, the mean detention time, and the temperature at which biomethanation is performed. The volumetric load, expressed in kg of organic matter as volatile solids (VS) or chemical oxygen demand (COD) introduced into the methane digester per m^3 of digester working volume and per day, takes into account the flow rate of the feed as well as the concentration (in VS or COD) of the influent.

The mean detention time, expressed in days (d), gives the average length of time during which the influent substrate biomass stays inside the methane digester. When there is no accumulation of solid particles, inter alia of active biomass in the methane digester, the mean detention time of the liquid (mean hydraulic detention time) is the same as for the solids. When there is accumulation of solid particles, inter alia of active biomass, in the methane digester as in the contact process, the anaerobic filter or the upflow system without carrier, the mean hydraulic detention time can be much shorter than the mean solids detention time.

The former definitions of volumetric load and mean detention time are only valid sensu stricto for continuous methane digestion systems, but can apply with appropriate changes also to batch methane digestion systems. For comparison purposes among processes, the volumetric load will correspond in batch methane digestion systems to the initial load divided by the number of days during which the batch methane digestion occurs, the latter corresponding to the mean detention time.

Finally, biomethanation can operate within three ranges of temperature : first the psychrophilic range which corresponds to ambient temperature and below, secondly the mesophilic range which begins at about 30 °C and ends at 40 °C, and, thirdly, the thermophilic range from 50 °C upwards.

III.1.1. Running conditions for the biomethanation of liquid and semi-solid manures

Table I.3.1 details the mean values of the volumetric loads encountered in the survey. The volumetric loads encountered for the full-scale biomethanation of liquid or semi-solid manures vary from 1 to 5.8 kg volatile solids introduced in the digester per m^3 of digester working volume per day in the case of one-step, once-through continuous biomethanation systems, and this holds for completely mixed, partially mixed and plug-flow systems.

Surprisingly, it appears that these values are generally not much below the optimal recommended value i.e. about 4 kg volatile solids introduced in the digester per m^3 of digester working volume per day. Indeed, the mean values elicited by the survey show that the volumetric loads applied for the biomethanation of manures are equal to or higher than 2 kg VS m^{-3} d^{-1}.

The mean concentrations of organic matter in the influent manures, expressed as volatile solids, vary from 15 to 90 kg volatile solids per m^3 of influent. As can be seen in Table I.3.2, pig manure influents present the lowest concentrations in volatile solids. It should be mentioned here that pig manure is less concentrated than cattle or poultry manure when it is produced. Moreover, as seen before in a number of cases e.g. in large piggeries, pig manure is often diluted by wash-waters and/or drinker leaks which again lower the concentration in volatile solids.

As regards the mean detention times in use for the biomethanation of the manure, Table I.3.3 shows that the mean values vary from 19 to 29 days, although the extreme mean detention times are either very short or very long. The very long mean detention times encountered in the survey indicate either an oversizing of the methane digester or a lack of availability of manure. On the other hand, when biomethanation occurs under psychrophilic conditions, mean detention times as long as 80-90 days are used. The mean hydraulic detention times applied in the pilot experiments on methane digestion of pig manure on anaerobic filters vary from 3 to 5.5 days.

Finally, if we consider the temperatures at which the biomethanation of manures is performed, it appears from the survey that 5 % of the methane digesters operate at psychrophilic conditions, i.e. at ambient temperature and below. These methane digesters are generally located underneath the stables so that their temperature depends upon that of the stables. Twelve per cent of the methane digesters are operating in the range from 25° to 30 °C whereas the majority of the methane digesters i.e. 64 % are working in the narrow range between 31° and 35 °C. Sixteen per cent are working in the range from 36° to 40 °C and the final 4 % are working under thermophilic conditions i.e. at temperatures higher than 50 °C.

III.1.2. Running conditions for the biomethanation of solid wastes

As regards the running conditions for the biomethanation of solid agricultural wastes, i.e. manure with bedding, crop residues or solid industrial wastes such as slaughter-house wastes, only a few values are available, at least for full-scale operations. Loading rates equal or equivalent to 2 to

Table I.3.1. <u>Volumetric loads observed for the biomethanation of different manures</u>

Type of feedstock	Mean volumetric loads : B_V (kg VS m^3 d^{-1})	Extreme values of volumetric loads	Number of cases
Dairy cattle manure	3.2	1.8 - 5.8	9
Bovine cattle manure	3.0	1.1 - 4.5	5
Pig manure	2.3	1.0 - 4.5	26
Poultry manure	1.9	1.8 - 2.0	2
Pig + dairy cattle manure	2.4	1.2 - 4.1	6
Bovine cattle + pig manure	2.9	1.9 - 3.9	3

Table I.3.2. <u>Mean concentrations in volatile solids of the different influent manures</u>

Type of feedstock	Mean concentration of the influent (kg VS m^{-3})	Extreme values of volatile solids concentrations	Number of cases
Dairy cattle manure	67	60 - 85	4
Bovine cattle manure	59	43 - 90	4
Pig manure	33	15 - 67	16
Poultry manure	57	45 - 70	2
Pig + dairy cattle manure	39	17 - 66	7

Witharen (The Netherlands) 1980. 195 m^3 completely mixed methane digester on a mixture of dairy cattle and pig manure.

Reading (United Kingdom) 1982. Twin 130 m^3 completely mixed methane digesters on dairy cattle manure.

Bettona (Italy) 1982. 7 000 + 2 500 m^3 completely mixed methane digester on pig manure.

Hagendorn (Switzerland) 1979. 285 m^3 completely mixed continuous methane digester on dairy cattle and pig manure.

Table I.3.3. Mean detention times observed for the biomethanation of different manures

Type of feedstock	Mean detention time (days)	Extreme values of mean detention time	Number of cases
Dairy cattle manure	29	14 - 67	12
Bovine cattle manure	28	11 - 83	27
Pig manure	19	4 - 45	47
Poultry manure	29	18 - 38	4
Pig + dairy cattle manure	22	10 - 35	6
Pig + bovine cattle manure	27	14 - 46	6

3 kg volatile solids introduced in the digester per m^3 of digestion working volume per day are mentioned both for batch systems and for one-step, once-through, continuous systems. The concentration of the influent manure is a function of the system of biomethanation applied. A concentration range of 160-200 kg of volatile solids per m^3 of influent is found for batch systems whereas for one-step, once-through continuous systems, a concentration range from 25 to 70 or even 100 kg volatile solids per m^3 of influent is found.

The mean detention times in use are once again a function of the system. They range from 28 to 246 days in the batch systems whereas they range from 17 to 46 days in the one-step, once-trough continuous systems. All these methane digesters on solid agricultural wastes are operating in the mesophilic range of temperature, generally between 30° and 38 °C, with one exception which works at 20°-23 °C.

III.1.3. Running conditions for the biomethanation of industrial wastewaters

The running conditions for the biomethanation of industrial wastewaters depend essentially on the type of biomethanation system involved. Table I.3.4 shows the different ranges of loading rates, concentrations of influent, mean hydraulic detention times, and temperatures applied for full-scale biomethanation.

As can be seen, the majority of the biomethanation systems involve accumulation of active biomass either by recirculation of active biomass as in the anaerobic contact and the IRIS system, or by trapping of the active

Table I.3.4. Biomethanation of wastewaters : running conditions in function of the different biomethanation systems

Type of feedstock	Type of biomethanation	Running conditions			
		Loading rates B_V kgCOD m^{-3} d^{-1}	Concentration of the influent kgCOD m^{-3}	Mean hydraulic detention time (days)	Temperature (°C)
Sugary waste-water	IRIS system	2 - 5.6	4.5 - 8.3	1 - 1.5	30 - 35
	Anaerobic contact	0.8 - 2.1	-*	3.6 - 8.3	30 - 35
	Upflow without carrier	12 - 16	-*	0.2 - 0.7	30 - 35
	Anaerobic filter	8.2	22 - 45	1.2	30 - 35
Sugary + distillery wastewater	IRIS system	8	-*	1.6	35 - 40
Distillery waste-water	IRIS system	4.2 - 7	14.1 - 38	3 - 5.5	35
	Anaerobic contact	2.8	22	8	50
	Completely mixed	2.2 - 2.5	20 - 23	9	35
Starch factory wastewater	Anaerobic contact	-*	-*	5	37
	Completely mixed	3.6	1.7	4.6	35 - 37
	Upflow without carrier	11	4 - 6	0.7	35
Potato wastewater	Completely mixed with recirculation of effluent	0.6	8.8	17	24
	Upflow without carrier	3.5 - 6	3 - 6	1.5	35
Vegetable canning wastewater	Anaerobic contact	2	5 - 10	5	35 ± 5
Pectinery waste-water	Anaerobic contact	1.7 - 5.3	7 - 23	3.8 - 5	37 - 38
Flax retting wastewater	Upflow without carrier	8 - 10	-*	0.4 - 0.7	35

* Not available

biomass inside the methane digester, without immobilization as in the upflow system without carrier, or with immobilization as in the anaerobic filter process. A consequence of this accumulation of biomass is that applied loading rates are higher than those in use for the one-step, once-through continuous systems. In particular, the upflow system without carrier and the anaerobic filter permit loading rates of about 10 kg chemical oxygen demand per m^3 digester working volume per day which are at least three times higher than those in use for the biomethanation systems without accumulation of active biomass.

The mean hydraulic detention times are also much shorter for the systems with accumulation of biomass. Mean hydraulic detention times of less than 1 day are currently observed in the upflow systems without carrier whereas mean hydraulic detention times of 1 to 5 or even 8 days are applied on anaerobic contact and IRIS systems. As a general rule, methane digesters treating industrial wastewaters operate under mesophilic conditions i.e. at temperatures ranging from 30° to 38 °C. Only one plant is operating at 24 °C and another in Greece operates at 50 °C.

III.2. Biogas productivities

Biogas productivities have been obtained in the survey for each biomethanation system and for each category of waste. The first parameter of biogas productivity is the gas production rate r_V, i.e., the number of m^3 of gas produced per m^3 of digester working volume per day. This gas production rate is a function of five factors : the type of waste, the biomethanation system, the volumetric loading rate, B_V, the mean detention time, and the temperature at which biomethanation occurs. The second parameter is the methane content of the biogas as volume per cent, which gives information on the quality of the gas. The third parameter of biogas productivity is the gas yield, i.e. the number of m^3 of biogas produced per kg of volatile solids or per kg of COD introduced in the methane digester. The values of gas yield permit a comparison of the biomethanation of different wastes for an identical system and in the same range of temperature. Finally, the fourth parameter is the conversion, i.e. the quantity of organic matter, expressed in volatile solids or in chemical oxygen demand, removed by the methane digester process per quantity of organic matter introduced in the methane digester. Conversion is essentially a measure of depollution.

Mean values for each of the biogas productivity parameters were calculated from the answers obtained in the survey. These values were classified according to the type of waste, the type of similar biomethanation system, and the range of temperature. Indications of the mean volumetric loading rate (B_V) in kg volatile solids introduced in the methane digester per m^3 of digester working volume per day and on the mean detention time (in days) are included for each classification (see Tables I.3.5.-I.3.9).

The biogas productivities will be successively examined for the three main types of wastes : liquid and semi-solid manures, solid (including manures with bedding and slaughter-house wastes) and industrial wastewaters. Within each category of waste, the influence of the biomethanation system, the temperature, and the nature of the waste are analysed.

III.2.1. Biogas productivities of liquid and semi-solid manures

Tables I.3.5-7 give the biogas productivities for the biomethanation of cattle manure, pig manure and mixed manures respectively.

a) Influence of the biomethanation system on biogas productivity

Liquid and semi-solid manures are mostly biomethanized in the one-step, once-through continuous methane digestion systems i.e. the completely mixed, the partially mixed and plug-flow systems, as noted in Part I, Chapter II. Other biomethanation systems such as the two-step continuous system, the batch system and even systems with accumulation of biomass have also been used for the methane digestion of these manures but to a lesser extent.

With the exception of the anaerobic filter system, no significant improvement of biogas productivities was observed when using any biomethanation system other than the most currently applied one-step, once-through continuous system. (At least, not enough answers were received which indicated improved or reduced biogas productivities). On the other hand, the anaerobic filter system, when applied to the biomethanation of pig manure, seems to yield promising results. Indeed, although the results presented in Table I.3.6 refer only to pilot-scale experience, it can be observed that, on the one hand, the gas yields obtained even at a lower temperature, i.e. 25 °C are identical to the ones obtained with the one-step, once-through continuous systems at 35 °C. On the other hand, since higher volumetric loading rates can be applied with the anaerobic filter systems, gas production rates four times faster than with the classical biomethanation systems are encountered. On the basis of these results and observations, the anaerobic filter system can be recommended for the biomethanation of the more liquid manures such as pig manure.

b) Influence of the nature of the manure

The gas productivity, expressed as gas yield, i.e. m^3 of biogas produced per kg of volatile solids introduced into the methane digester permits a comparison between the different manures. When comparing the gas yields obtained with the same type of biomethanation system, i.e. the one-step, once-through continuous system, for the same range of temperature varying from 31° to 35 °C, the lowest gas yield is observed for dairy cattle manure, i.e. 0.21 m^3 of biogas produced per kg of volatile solids introduced in the methane digester (Table I.3.5). A gas yield of 0.35 m^3 of biogas produced per kg of volatile solids introduced in the digester is obtained for bovine cattle manure whereas the highest gas yield is observed for pig manure, i.e. a gas yield of 0.45 m^3 of biogas produced per kg of volatile solids introduced in the digester. Biomethanation of pig manure seems also to yield a better quality of gas than the biomethanation of cattle manure.

The gas productivities obtained with pig manure are confirmed by the calculated conversion values. Indeed, a mean conversion of 0.54 kg volatile solids removed by biomethanation per kg volatile solids introduced in the digester is obtained for pig manure, whereas the comparative figures obtained for dairy cattle manure and bovine cattle manure are 0.17 and 0.38 kg respectively. As can be observed in Table I.3.7., upon examination

Table I.3.5. Biogas productivities from cattle manure

Volumetric loading rate : B_V ($kg\ VS\ m^{-3}\ d^{-1}$)	Detention time (days)	Temperature (°C)	Gas production rate : r_V ($m^3\ m^{-3}\ d^{-1}$)	Methane content (%)	Gas yield : $Y_{gas/VSo}$ ($m^3\ kg^{-1}$)	Conversion : $Y_{Sr/So}$ ($kg\ kg^{-1}$) COD	VS
			a) Dairy cattle manure				
- One-step, once-through continuous system							
1.8 (1)*	33 (1)	26	0.23 (1)		0.13 (1)		0.55 (1)
1.8 (1)	39 (1)	30	0.36 (1)		0.21 (1)	0.24 (1)	0.17 (1)
2.4 (3)	33 (3)	31 - 35	0.47 (3)	63 (3)	0.21 (3)		0.17 (2)
4.7 (3)	18 (3)	36 - 37	1.08 (3)	63 (2)	0.27 (3)	0.33 (1)	0.33 (1)
	10 (1)	57	1.50 (1)	63 (1)			
- Anaerobic contact							
4.3 (1)	15 (1)	35	1.44 (1)	62 (1)	0.33 (1)		
			b) Bovine cattle manure				
- One-step, once-through continuous system							
	20 (1)	30	0.75 (1)	60 (1)			
2.3 (8)	29 (26)	31 - 35	0.89 (23)	61 (20)	0.35 (9)		0.38 (8)
	17 (3)	36 - 40	1.45 (3)	63 (3)			
- Two-step, once-through continuous system							
2.5 (1)		35	1.30 (1)	67 (1)	0.51 (1)		
- Upflow system without carrier							
2.1 (2)	16 (2)	34	0.81 (2)	62 (2)	0.37 (2)		

* Figures in brackets indicate the number of cases analysed

Table I.3.6. Biogas productivities from pig manure

Volumetric loading rate, B_V ($kg\ VS\ m^{-3}\ d^{-1}$)	Detention time (days)	Temperature (°C)	Gas production rate : r_V ($m^3\ m^{-3}\ d^{-1}$)	Methane content (%)	Gas yield : $Y_{gas/So}$ ($m^3\ kg^{-1}$) VS	COD	Conversion : $Y_{Sr/So}$ ($kg\ kg^{-1}$) COD	VS
- One-step, once-through continuous system								
- *	90 (1)**	20	0.2 (2)	50 (1)				
- *	80 (1)	20 - 25	0.4 (2)	67 (1)				
2.5 (2)	26 (3)	26 - 30	0.6 (3)	71 (3)	0.29 (2)	0.24 (1)	0.37 (1)	0.22 (1)
2.2 (16)	19 (21)	31 - 35	1.1 (24)	67 (15)	0.45 (15)		0.50 (4)	0.54 (9)
2.1 (7)	16 (7)	36 - 40	1.0 (7)	72 (7)	0.46 (7)		0.57 (4)	0.38 (4)
- Anaerobic filter								
4.2 (COD) (1)	5 (1)	25	1.1 (1)	86 (1)		0.26 (1)	0.66 (1)	
15.0 (COD) (1)	3 (1)	29	3.5 (1)	80 (1)		0.35 (1)	0.31 (1)	0.39 (1)
8.4 (1)	3 (1)	35	4.8 (1)	80 (1)	0.34 (1)		0.70 (1)	
5.8 (1)	4 (1)	38	1.6 (1)	80 (1)	0.28 (1)			
- Two-step, once-through continuous system								
4.1 (3)	14 (3)	35	1.2 (3)	73 (2)	0.34 (3)		0.45 (2)	0.55 (1)

* Not available
** Figures in brackets indicate the number of cases analysed

Table I.3.7. Biogas productivities from mixed manure

Volumetric loading rate : B_V (kg VS m^{-3} d^{-1})	Detention time (days)	Temperature (°C)	Gas production rate : r_V (m^3 m^{-3} d^{-1})	Methane content (%)	Gas yield : $Y_{gas/VSo}$ (m^3 kg^{-1})	Conversion : $Y_{Sr/So}$ (kg kg^{-1}) VS	Conversion : $Y_{Sr/So}$ (kg kg^{-1}) COD
a) Pig + dairy cattle manure							
- One-step, once-through continuous system							
1.4 (2)*	30 (2)	30	0.39 (2)	64 (1)	0.29 (2)		0.25 (2)
3.0 (3)	16 (3)	31 - 35	0.83 (3)	67 (3)	0.32 (3)	0.47 (2)	0.41 (2)
- Two-step, once-through continuous system							
2.5 (1)	25 (1)	35	0.86 (1)	62 (1)	0.34 (1)		
- Batch							
1.1 (1)	35 - 95 (1)	35	0.18 (1)		0.16 (1)		
b) Pig + bovine cattle manure							
- One-step, once-through continuous system							
	46 (1)	30	0.65 (1)	73 (1)			
1.7 (1)	27 (3)	31 - 35	1.00 (3)	52 (2)	0.28 (1)		
3.3 (1)	23 (1)	37	0.78 (1)		0.23 (1)		

* Figures in brackets indicate the number of cases analysed.

of the biogas productivities for mixed manures, no conclusion can be drawn pointing to possible synergetic effects during the biomethanation of pig manure mixed with cattle manure.

c) Influence of temperature

It can be observed from Tables I.3.5, I.3.6 and I.3.7 that the biogas production rate per unit digester working volume increases when the temperature increases. There are two reasons for this : first at increased temperatures, shorter mean detention times are usually applied with the result that the volumetric loading rate is increased. On the other hand, there appears to be an increase of the gas yield expressed as m^3 of biogas produced per kg of volatile solids introduced in the methane digester upon raising the temperature. Nevertheless, the increase in gas yield is not proportional to the increase in temperature. As can be observed in Fig. I.3.1 which relates to the gas yields of pig manure biomethanized in the one-step, once-through continuous systems, gas yields remain nearly identical in the range of temperatures from 30° to 35 °C and from 35° to 40 °C, whereas gas yields are one-third lower for the range of temperatures from 25° to 30 °C. There were unfortunately too few results to confirm this observation conclusively.

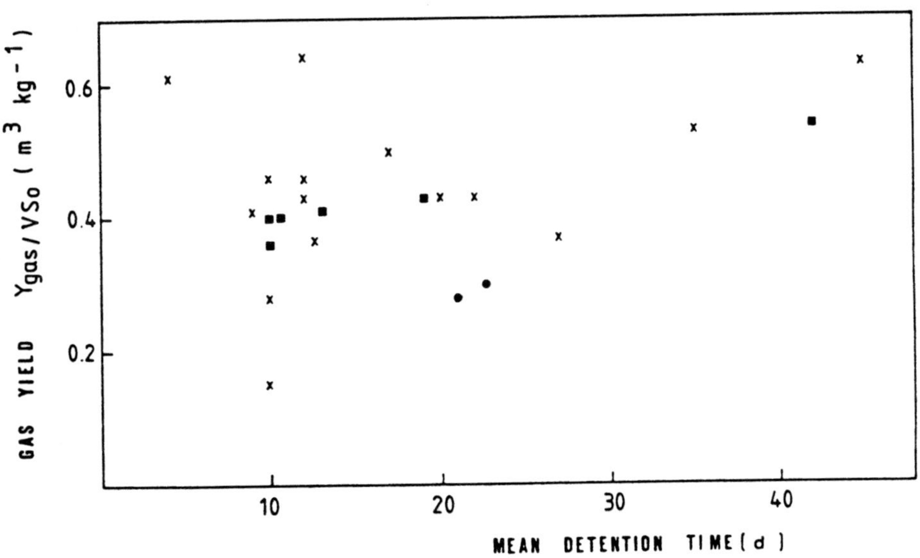

Fig. I.3.1. Evolution of biogas yields as a function of the mean detention time at different temperatures in the biomethanation of pig manure in the one-step, once-through continuous methane digestion systems

● : Temperatures between 25° and 30 °C; x : Temperatures between 30° and 35 °C; ■ : Temperatures between 35° and 40 °C.

III.2.2. Biogas productivities of solid wastes

The biogas productivities of the solid agricultural wastes, i.e. dairy and bovine cattle manure with bedding and of industrial solid wastes, i.e. slaughter-house wastes, are presented in Table I.3.8. For the biomethanation of these solid wastes, two systems may be compared : the one-step, once-through continuous system and the batch system. The former continuous system yields higher gas production rates and seemingly higher gas yields than the batch system. When comparing the gas productivities of manure with and without bedding, it appears that the gas yields are nearly the same, at least in the case of one-step, once-through continuous systems. Slaughter-house waste produces a higher gas yield than manure with bedding. Not enough results on solid wastes have been obtained for a more detailed analysis.

III.2.3. Biogas productivities of industrial wastewaters

Table I.3.9 gives the biogas productivities for the biomethanation of wastewaters as a function, first, of the nature of the wastewater and, secondly, as a function of the biomethanation system.

a) Influence of the biomethanation system

Unlike the biomethanation of liquid and semi-solid manures, industrial wastewaters are mostly biomethanized in systems with accumulation of active biomass, i.e. the anaerobic contact and the IRIS process, the anaerobic filter, and the upflow system without carrier. These different biomethanation systems have been described in the introductory chapter. The case of biomethanation of sugary wastewaters can be taken for the purpose of comparison between these different systems. Fig. I.3.2. shows the different gas yields expressed in m^3 of biogas produced per kg COD introduced in the methane digestion system as a function of four different biomethanation systems. It can be observed that the upflow system without carrier produces the same biogas yields as are observed for the other biomethanation systems but with much shorter mean hydraulic detention times : up to 10 times shorter when compared with those obtained in the anaerobic contact system, or 2-3 times shorter when compared with those obtained in the anaerobic filter or the IRIS systems. The upflow system without carrier consequently yields very high gas production rates expressed in m^3 of biogas produced per m^3 of digester working volume per day : 5.4 m^3 m^{-3} d^{-1}, compared with 1 or 2 m^3 m^{-3} d^{-1} for the other systems. Nevertheless, it can be observed in Table I.3.9 that, nothwithstanding these high gas production rates, the conversions, expressed as kg COD removed by biomethanation per kg COD introduced in the digestion system are maintained.

The upflow system without carrier also yields gas productivities better than those obtained with the other biomethanation systems for wastewaters other than sugary wastewaters, although the differences are not so spectacular. Indeed, the nature of the wastewaters plays an important role in biogas productivities.

Table I.3.8. Biogas productivities from solid agricultural and industrial wastes

Volumetric loading rate : B_V ($kg\ VS\ m^{-3}\ d^{-1}$)	Detention time (days)	Temperature (°C)	Gas production rate : r_V ($m^3\ m^{-3}\ d^{-1}$)	Methane content (%)	Gas yield : $Y_{gas/VSo}$ ($m^3\ kg^{-1}$)	Conversion : $Y_{Sr/So}$ ($kg\ kg^{-1}$) COD	VS
a) Dairy cattle manure with bedding							
- One-step, once-through continuous system							
5.3 (2)*	22 (3)	36 (3)	1.14 (3)	60 (2)	0.21 (2)	0.52 (1)	0.73 (1)
- Two-step, anaerobic filter							
4.0 (1)	20 (1)	37 (1)	0.51 (1)	62 (1)	0.13 (1)	0.62 (1)	0.80 (1)
- Batch							
	28 (1)	20-23 (1)	0.44 (1)				
b) Bovine cattle manure with bedding							
- One-step, once-through continuous system							
3.7 (1)	30 (2) 25 (1)	35 38	1.10 (2) 1.20 (1)	66 (2) 70 (1)	0.30 (1)		
- Batch							
3.0 (1)	56 (1)	35	0.50 (1)	54 (1)	0.17 (1)		0.36 (1)
c) Slaughter-house wastes							
- One-step, once-through continuous system							
2.3 (2)	32 (2)	35	1.17 (2)	65 (2)	0.54 (2)	0.34 (2)	0.45 (1)
- Batch							
3.0 (1)	42 (1)	35	0.65 (1)	67 (1)	0.20 (1)		

* Figures in brackets indicate the number of cases analysed.

Table I.3.9. Biogas productivities from wastewaters

Volumetric loading rate : B_V (kg COD m^{-3} d^{-1})	Detention time (days)	Temperature (°C)	Gas production rate : r_V (m^3 m^{-3} d^{-1})	Methane content (%)	Gas yield : $Y_{gas/CODo}$ (m^3 kg^{-1})	Conversion : $Y_{CODr/CODo}$ (kg kg^{-1})
			a) Sugary wastewaters			
Anaerobic contact						
1.6 (3)*	5.3 (3)	35 (3)	0.64 (3)	75 (3)	0.40 (3)	0.89 (3)
IRIS system						
4.7 (3)	1.5 (3)	35 (3)	0.88 (3)	73 (3)	0.26 (3)	0.91 (3)
Upflow without carrier						
13.7 (5)	0.5 (6)	35	5.4 (6)	75 (6)	0.41 (6)	0.86 (3)
Anaerobic filter						
8.2	1.2 (1)	35	2.0 (1)	70 (1)	0.24 (1)	0.70 (1)
			b) Distillery wastewaters			
IRIS system						
5.6	4.3 (2)	35 (1)	2.53 (2)	59 (2)	0.42 (1)	0.83 (2)
Completely mixed system						
3.8 (1)	10	35 (1)	1.50	65 (1)	0.40 (1)	
Upflow without carrier (two-step; pilot-scale)						
20.3 (1)	1.25(1)	32.5 (1)	6.1	67 (1)	0.29 (1)	0.67 (1)

* Figures in brackets indicate the number of cases analysed.

Table 1.3.9. Biogas productivities from wastewaters (continuation)

Volumetric loading rate : B_V (kg COD m^{-3} d^{-1})	Detention time (days)	Temperature (°C)	Gas production rate : r_V (m^3 m^{-3} d^{-1})	Methane content (%)	Gas yield : $Y_{gas/CODo}$ (m^3 kg^{-1})	Conversion : $Y_{CODr/CODo}$ (kg kg^{-1})
c) Sugary + distillery wastewater						
- IRIS system						
8.0	1.6	35 - 40	3.3 (1)*	60 (1)	0.40 (1)	
d) Starch factory wastewater						
- Anaerobic contact						
-**	5 (1)	37 (1)	1.1 (1)	65 (1)		
- Completely mixed system						
3.6	4.6	36	1.1 (1)		0.31 (1)	
- Upflow without carrier						
11 (1)	0.7 (1)	35 (1)	3.0 (1)	70 (1)	0.27 (1)	0.85 (1)
e) Potato processing wastewater						
- Completely mixed with recirculation of effluent						
0.6	17	24	0.3	60	0.47	0.84
- Upflow without carrier						
4.8 (2)	1.5 (1)	35	2.0 (2)	76 (2)	0.40 (2)	0.83 (2)

* Figures in brackets indicate the number of cases analysed
** Not available.

Table I.3.9. Biogas productivities from wastewaters (end)

Volumetric loading rate : B_V ($kg\ COD\ m^{-3}\ d^{-1}$)	Detention time (days)	Temperature (°C)	Gas production rate : r_V ($m^3\ m^{-3}\ d^{-1}$)	Methane content (%)	Gas yield : $Y_{gas/CODo}$ ($m^3\ kg^{-1}$)	Conversion : $Y_{CODr/CODo}$ ($kg\ kg^{-1}$)
			f) Pectinery wastewater			
- Anaerobic contact						
2.9 (2)*	3.4 (2)	37 (2)	1.1 (2)	66 (2)	0.36 (1)	0.79 (1)
			g) Vegetable canning wastewater			
- Anaerobic contact						
2.0 (1)	5 (1)	35 ± 5	1.0 (1)	60 (1)	0.45 (1)	0.92 (1)
			h) Flax retting wastewater			
- Upflow without carrier						
9.0 (1)	0.6 (1)	35	4.0 (1)	67 (1)	0.45 (1)	0.87 (1)
			i) Dairy wastewater			
- Anaerobic filter						
5.0 (1)	0.5 (1)	28 (1)	1.8 (1)	82 (1)	0.35 (1)	0.82 (1)

* Figures in brackets indicate the number of cases analysed.

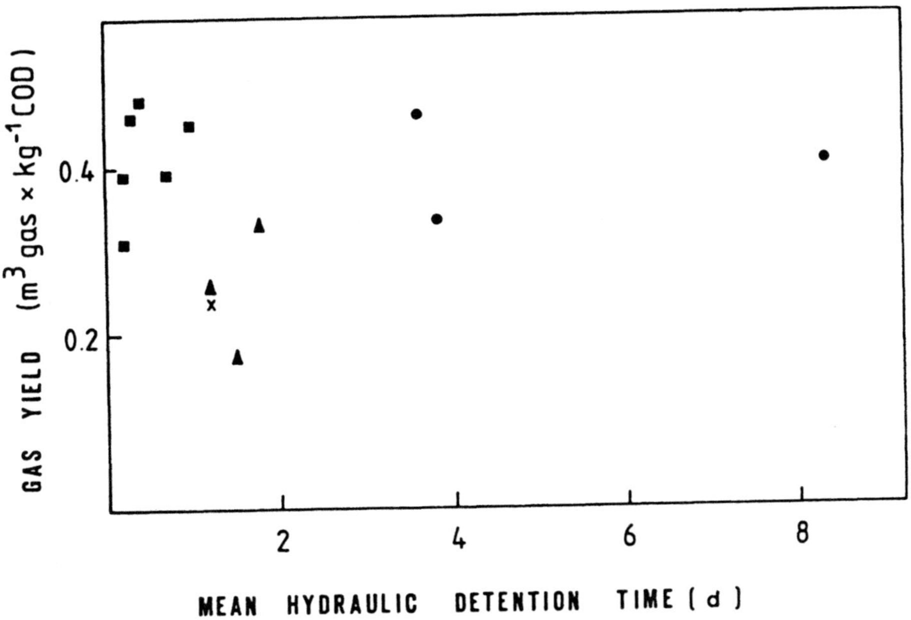

Fig. I.3.2. Gas yields observed for sugary wastewaters as a function of the mean hydraulic detention time for different biomethanation systems

■ : Upflow system without carrier; ▲ : IRIS system; ● : Anaerobic contact; X : Anaerobic filter.

b) Influence of the nature of wastewaters

Whereas the biogas production rates depend essentially on the system of biomethanation, the gas yields, i.e. the number of m^3 of biogas produced per kg COD introduced in the digestion system depend essentially on the quality of the wastewaters. Even within the same type of industry, the quality of the wastewaters to be treated changes. High-strength wastewaters may contain either important concentrations of highly polymerized compounds or important concentrations of simple molecules. The former are generally less biodegradable, both in yield and rate, than the latter. Therefore, the gas yields as m^3 of biogas obtained per kg COD in the load are lower with the former than with the latter. As an example, in Table I.3.9, it can be observed that starch factory wastewaters as well as certain sugary and distillery wastewaters present gas yields below 0.30 m^3 of biogas produced per kg COD added as load, whereas gas yields above 0.35 m^3 of biogas produced per kg COD added as load are obtained for most of the other wastewaters.

Genappe (Belgium) 1980. Twin 1 530 m^3 IRIS methane digesters on sugary wastewater.

Table I.3.9 also shows that the conversions obtained, expressed in kg COD removed by biomethanation per kg COD added in the methane digestion system are generally high, irrespective of the type of process applied and of the type of wastewater. With the exception of three cases, the conversions observed are higher than 0.8 kg COD removed by biomethanation per kg COD added as load. Hence it can be concluded that biomethanation is an attractive process for depolluting industrial wastewaters.

c) Influence of temperature

Most wastewaters are treated under mesophilic conditions. In practice, the wastewaters are either still warm when they arrive at the biogas plant, or they are heated by using excess energy produced in the factory. Hence it is rare to treat wastewaters at temperatures lower than 30 °C. Two cases of biomethanation at 24 °C and 28 °C respectively were encountered : for the treatment of potato processing wastewater and dairy wastewater. Consequently the influence of temperature on the biomethanation of industrial wastewaters can not yet be analysed.

ANAGNI (Italy) 1982. 6 000 + 2 500 m^3 completely mixed methane digesters on distillery wastewater.

CHAPTER IV. ECONOMICS

IV.1. Introduction

The survey produced limited information on the investment cost of agricultural and industrial biogas plants. All figures obtained and presented here include the cost of the complete biogas plant, i.e. the methane digester itself, the system of gas storage and utilization, and the systems of influent and effluent storage and/or treatment, if any. In order to compare the investment cost of the different biogas plants, the total cost was first divided by the number of m^3 working volume of the methane digester and, secondly, this figure per m^3 was converted into European Currency Units (ECU). The exchange rates employed for the different European currencies are indicated below(1). The price per m^3 expressed in ECU was then updated to end 1982.

The second part of this chapter is devoted to the calculation of pay-back periods of biogas plants. Little information is as yet available for such an economic analysis. Indeed, of the biogas plants existing in the countries covered by the survey at the end of 1982, only a few plants had been in routine operation for more than one year, and if so, their economics had not always been properly evaluated. Nevertheless, valuable data were obtained for 32 biogas plants on farms.

IV.2. Investment cost of biogas plants

IV.2.1. Biogas plants on farms

The investment cost of agricultural biogas plants may be classified in four categories, depending on whether the plant was or was not constructed by the farmers themselves and on whether there is or is not an engine included in the biogas plant. The mean investment cost of agricultural biogas plants is presented in Table I.4.1. The mean values obtained are first presented independently of the countries and secondly by country.

Biogas plants constructed by the owner cost on average 26 % less than plants constructed by a company. The difference can be explained by the fact that the cost of labour is not always included in the first case; moreover, the owner often uses materials available on site for the construction. The integration of an engine for generating electricity increases the investment cost by 12 % but of course brings all the advantages of an autonomous source of electrical energy.

---

(1) Values of the ECU in the different European currencies (May 1982) : 45.0106 BFR; 8.07375 DKR; 2.38745 DM; 64.4975 DRA; 6.21548 FF; 0.690075 IRL; 1325.76 LIT; 2.64895 HFL; 0.568477 UKL; 1.99134 SFR.

Table I.4.1. Mean investment cost of agricultural plants. Updated prices expressed in ECU/m³ digester working volume

	Construction by the owner, without engine	Construction by the owner, with engine	Construction by a company, without engine	Construction by a company, with engine
	1. Mean values independently of the country			
	353 (31)(1)	416 (6)	479 (44)	555 (39)
	2. Mean values per country			
Belgium	372 (5)	289 (1)	447 (5)	460 (2)
Denmark	168 (2)	103 (1)	412 (1)	481 (5)
France	338 (9)	1 116 (1)	476 (2)	625 (4)
F.R. Germany	417 (12)	447 (2)	648 (19)	952 (7)
Ireland			251 (1)	
Italy			293 (3)	545 (9)
The Netherlands	142 (1)	93 (1)	309 (10)	315 (9)
United Kingdom	268 (2)		324 (3)	469 (3)
Switzerland			600 (*)	800 (*)

* Not available
(1) Figures in brackets indicate the number of answers considered.

It will be seen from Table I.4.1 that the investment cost fluctuates widely from country to country. The biogas plants of the Federal Republic of Germany appear in general to involve a higher investment cost than those of the other countries. It must however be added that for the "do it yourself" German digesters, the manpower of the farmer was taken into account at a rate of 12 DM/hour. This was not the case for the other countries.

Investment costs vary also with the type of methane digester. For example, the fed batch psychrophilic methane digesters, for which transformed manure storage tanks serve as reaction tank, cost less than 100 ECU/m³ digester working volume, whereas more sophisticated methane digesters with more than one mixing device and additional storage or treatment tanks for the influent and effluent can cost over 500 ECU/m³ digester working volume.

When the investment costs are analysed as a function of the scale of the methane digester, it appears that the price per m³ digester working volume decreases with increasing scale up to 100 m³, when the investment cost per m³ becomes more or less stable. Figs. I.4.1, I.4.2 and I.4.3 show this relation for the different categories of biogas plants considered.

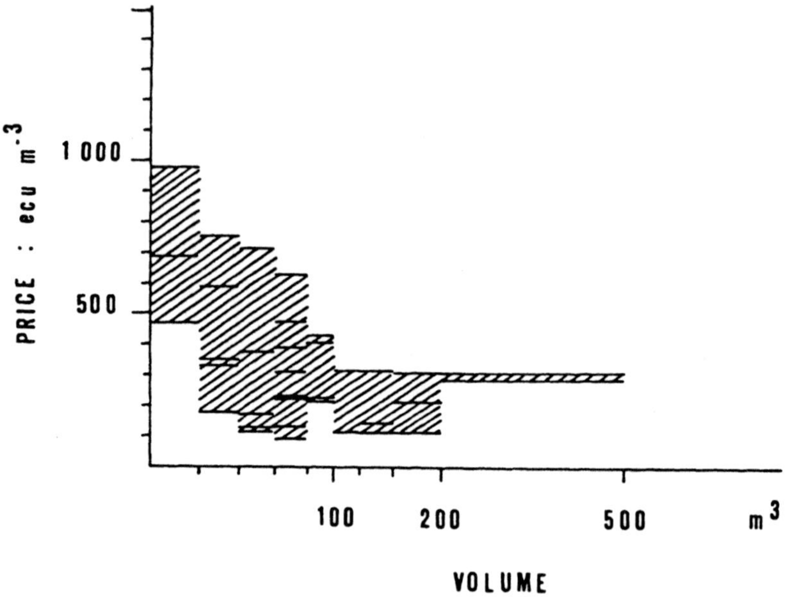

Fig. I.4.1. Evolution of the investment cost with the scale of the methane digester (constructed by owner, without engine)

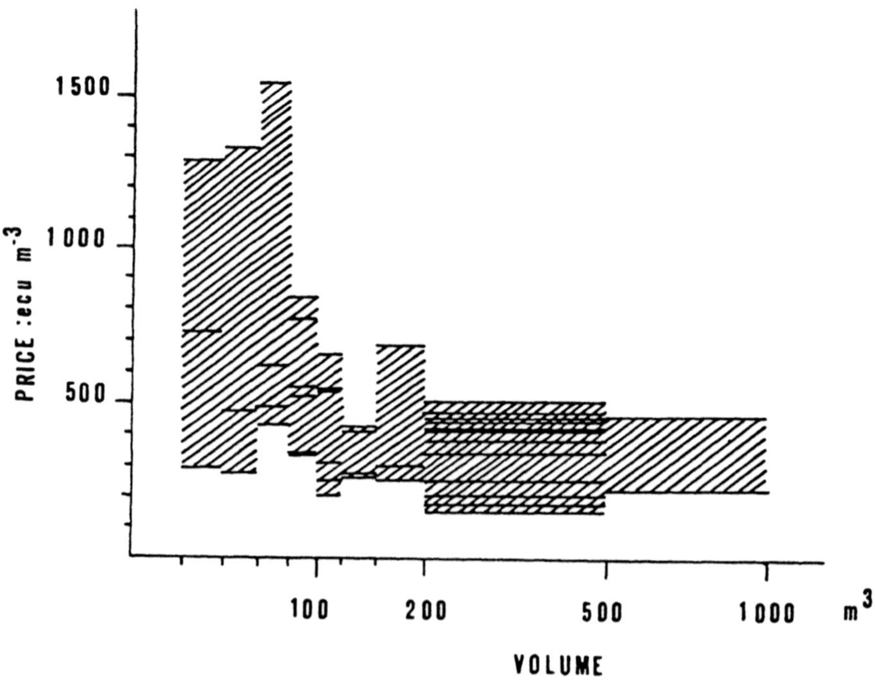

Fig. I.4.2. Evolution of the investment cost with the scale of the methane digester (constructed by a company, without engine)

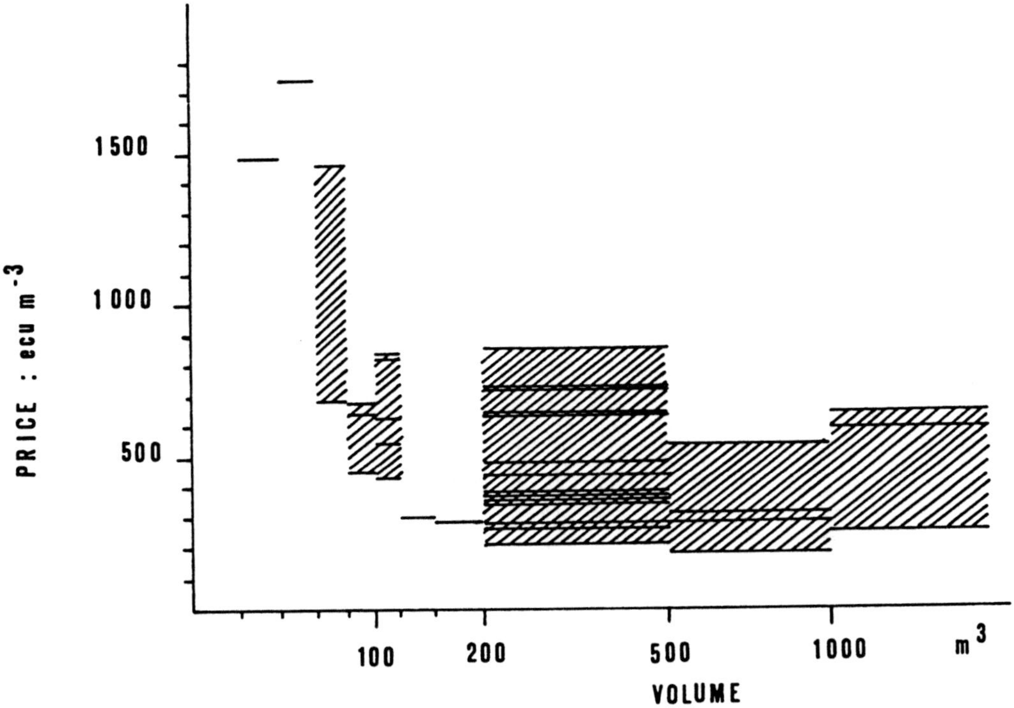

Fig. I.4.3. Evolution of the investment cost with the scale of the methane digester (constructed by a company, with engine)

IV.2.2. Biogas plants in industry

In order to properly compare the investment costs of industrial biogas plants, it is necessary to take into consideration the categories of biomethanation systems. Three categories are considered : the anaerobic contact system (completely mixed continuous system with recycle of active biomass after settling), the French IRIS (Institut de la Recherche de l'Industrie Sucrière) system (anaerobic contact in one tank), and the upflow system without carrier (upflow, non-mixed digester with accumulation of settled active biomass). Depending upon the biomethanation system, the gas production rate varies from about 1 to about 5 m^3 biogas per m^3 of digester working volume per day. As an example, for the biomethanation of sugary wastewaters, a mean gas production rate of 5.4 m^3 m^{-3} d^{-1} is obtained with the upflow system without carrier, whereas 0.64 and 0.88 m^3 m^{-3} d^{-1} are obtained respectively for the anaerobic contact and the IRIS systems. The investment cost per m^3 of digester working volume varies also in the same direction. A mean investment cost of 2 159 ECU m^{-3} digester working volume is observed for the upflow system without carrier, whereas mean investment costs of 436 and 248 ECU m^{-3} are observed respectively for the anaerobic contact and IRIS systems. Fig. I.4.4 depicts these observations.

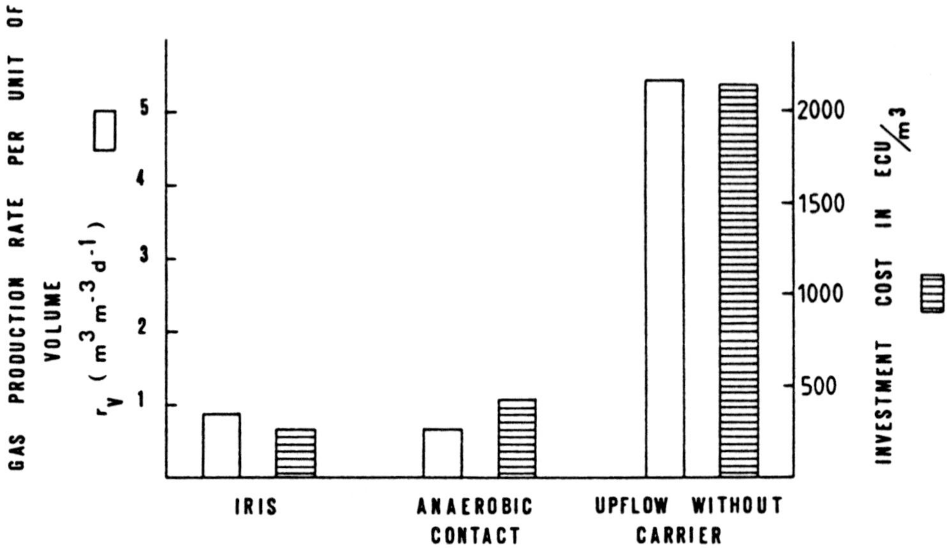

Fig. I.4.4. Investment cost/m^3 related to gas production rate for three biomethanation systems

It may be concluded that the relation between investment cost and performance is much the same for the three biomethanation systems. The IRIS system appears to be a little less expensive. The principal reason for this is that IRIS methane digesters have hitherto been constructed entirely or partly in polymeric material instead of steel or concrete which of course leads to lower investment cost. The tendency, however, is to construct the new IRIS methane digesters in more durable material so that finally their investment cost will be similar to that of the anaerobic contact methane digester.

The cost of the reactor itself in an agricultural biogas plant amounts usually to around 30 to 40 % of the overall cost. The cost of the reactor itself in an industrial biogas plant amounts usually to around 20 % of the overall cost. The reason for this difference is that industrial requirements for efficiency and reliability are more severe and hence require extra ancillary equipment. Often also industrial quotations include the costs of start up and one year follow up. These arguments are put forward by constructors of UASB biogas plants in order to explain their apparently higher specific investment prices.

As was the case for agricultural biogas plants, the investment cost is reduced when the industry constructs its own plant. However, the reduction is more important than for biogas plants on farms : for an anaerobic contact plant the investment cost decreases in this way from 436 to 152 ECU m^{-3}, a more than twofold reduction.

IV.3. Operation cost of biogas plants

The operation cost, including maintenance, of a biogas plant represents per year an average of 2 % of the investment cost (see below, Table I.4.3). Furthermore, it appears that the operation costs of plants with or without electricity generation are the same.

The mean amount of labour needed for the maintenance and operation of a biogas plant is about 166 hours per year, or about 30 minutes per day. Although few data were obtained, it seems that a digester which is operating well needs no more than 15 minutes' labour per day. Extra labour is needed for the cleaning of the heater or the engine and also for any repairs.

IV.4. Pay-back periods of biogas plants

Because the biogas plants on farms are the oldest existing biogas plants, complete and valid economic data exist for some of them and were obtained through the survey. Hence, 32 biogas plants could be analysed by an economic model.

IV.4.1. Description of the model

For this economic analysis, the Financial Model Processor, perfected as a processing tool by Schepens (Facultés Notre-Dame de la Paix, Namur, Belgium) was used. This model allows i.a. the calculation of the Simple Pay-Back Period, the Net Present Value (NPV) and the Internal Rate of Return (IRR) for each biogas plant. The Simple Pay-Back Period is the number of years it takes for a project to recover its original investment from the expected net cash-flows. The investment taken into account in the model is the investment of the whole biogas plant, including the methane digester itself, the storage tanks, if newly built, and the installation for biogas utilization, including manpower except for "do it yourself" biogas plants. The concept of Pay-Back Period does not take into account the cash-flows generated after the period required for repayment of the investment. Moreover, the Simple Pay-Back Period does not allow for the fact that one ECU cashed today is more valuable than one ECU cashed in the future.

On the contrary, the Net Present Value (NPV) and Internal Rate of Return take into consideration the fact that money cashed immediately is more valuable than money to be cashed at some future date. Indeed, the Internal Rate of Return (IRR) is defined as the discount rate that equates the present value of the expected net future cash-flows or receipts to the initial cash outlay. If the IRR is larger than the capital cost (k), the cash-flows generated by the project will repay the investment cost, including its interest whether due or not. The project can then be accepted. For this economic analysis, a reference capital cost, k, of 15 % was assumed. This value was chosen rather high so as to allow a factor of security for temporary poor operation of the methane digester, and so as to take into account the sometimes approximate character of the economic data provided by the owners of the plants.

In this model, it has been assumed that the life-span of a biogas plant could attain a mean period of 14 years, considering that the life-span of the reaction tank alone can attain 20-25 years and that the life-span of the mechanical part of the biogas plant can attain 5-8 years. Of course, these 14 years of life-span of a biogas plant are only an assumption and will certainly vary from plant to plant. It can be supposed also that expensive biogas plants will have a longer life-span than inexpensive ones, but this remains to be demonstrated in the future. Nevertheless, the choice of 14 years does not greatly influence the calculation of the Simple Pay-Back Period. However, the Internal Rate of Return does change with the assumed life-span. If, for example, a life-span of 10 years had been assumed, the Internal Rate of Return would be lower than it is for 14 years.

The inflation rates considered in this model are taken from "Economie Européenne : informations économiques : prévision pour 1982-83", supplement to number 10 of October 1982[1]. Table I.4.2 presents the different inflation rates per country.

Table I.4.2. Inflation rates per country

Country	Inflation rates (%)[1]
Belgium	8.5
Denmark	7
France	8.9
F.R.Germany	5.6
Italy	15.0
The Netherlands	4.5
United Kingdom	6.9[2]
Switzerland	5.3[3]

(1) Assumed rates for the next 14 years. From reference in footnote[1]
(2) Value for 1982. For the previous years, the value was 13.
(3) According to Wellinger.

In this economic analysis, the yearly expected net cash-flows were calculated directly by subtracting the yearly operation cost, including maintenance and repairs (for "do it yourself" methane digesters, the cost of manpower supplied by the farmer is again not taken into account) and the taxes from the yearly returns which are provided by the saving of energy costs at the farm due to autonomous biogas or electricity production, and the eventual sale of gas or electricity to third parties.

(1) Issued by the Commission of the European Communities, Directorate-General II : Economic and Financial Affairs.

IV.4.2. Results of the economic analysis

Of the 32 biogas plants analysed, 5 are "do it yourself" plants. Table I.4.3 summarizes the data and the economic results for these 5 biogas plants. The digester number indicated above each column refers to the listing presented in Appendix III of this book.

As can be observed in Table I.4.3, 3 of the 5 "do it yourself" biogas plants analysed are clearly profitable. Their Simple Pay-Back Period is 3 or 4 years and their Internal Rate of Return is higher than 30 %, which in turn is higher than the capital cost of 15 %. Essential for their profitability is their low investment cost. Indeed, although they have a low daily biogas production, less than 1 m^3 per m^3 of digester working volume, they are still profitable because their investment cost lies between 100 and 160 ECU/m^3.

Biogas plant 182, more costly for investment, is nearly profitable with 7 years of Simple Pay-Back Period. Finally, plant 238 is uneconomic with a Simple Pay-Back Period of 10 years and an Internal Rate of Return of 3.8 %. The latter plant is not costly for investment, but it operates only 6 months per year so that the savings obtained from the biogas production do not permit repayment of the investment cost.

The remaining 27 biogas plants included in the economic analysis were constructed by companies. Table I.4.4 summarizes the data and the economic results of 15 biogas plants without electricity generation devices constructed by companies, whereas Table I.4.5 concerns 12 biogas plants with electricity generation devices constructed by companies.

Of these 27 biogas plants, only 3 (nos. 150, 279 and 168) are profitable with Simple Pay-Back Periods of 5-6 years and Internal Rates of Return higher than 15 %. Six of the 27 plants (nos. 281, 339, 240, 6, 31 and 343) have Simple Pay-Back Periods of 8 to 9 years but an Internal Rate of Return lower than 15 %. The 18 remaining plants are clearly uneconomic with Simple Pay-Back Periods equal to, or higher than 10 years and Internal Rates of Return approaching zero or even negative.

The main reason for the non-profitability of 24 biogas plants varies from one plant to another. A synopsis is given in Table I.4.6. Two main reasons predominate : first, poor operation of the methane digester, and secondly, too high an investment cost. The three remaining reasons (poor gas utilization, important requirements for the heating of the digester, high operation cost) concern particular cases.

Table I.4.6. Main reasons for non-profitability of biogas plants

Reason	Share of total (%)
- Poor operation of the methane digester	46
- Investment cost too high	42
- Poor gas utilization	4
- Heating requirements too important	4
- Operation cost too high	4
	100

Table I.4.3. Economic analysis of "do it yourself" biogas plants (without electricity generation)

	Digester 145 (B)	Digester 4 (DK)	Digester 182 (F)	Digester 238 (NL)	Digester 390 (UK)
a. Location of the plant	Voormezele	Guldager	Montalieu	Garderen	Haverfordwest
Type of waste	Pig manure	Dairy cattle manure	Pig manure	Pig manure	Dairy cattle + pig manure + crop residues
Remarks				Operates 6 months per year	
Year of start-up	1981	1980	1980	1978	1979
Working volume (m^3)	73	44	3 x 30	300	66
b. Gas production rate ($m^3\ m^{-3}\ d^{-1}$)	1981 : 0.48 Thereafter : 0.72	0.68	0.66	1978–81 : 0.17 Thereafter : 0.27	0.2 - 1.7
c. Heating requirements (as % of b.)	30	60	25	30	30 - 50 Not in summer
d. Investment cost (ECU/m^3)	122	106	322	119	160
e. Yearly returns versus investment cost	1981 : 0.14 Thereafter : 0.21	0.27	0.11	1978–81 : 0.07 Thereafter : 0.11	0.22
f. Yearly operation cost versus investment cost	0.01	0.03	0.01	0.02	0.04
g. Pay-Back Period (years)	4	3	7	10	4
h. Internal Rate of Return (%)	32.3	39.8	13.7	3.8	30.4

Table I.4.4. Economic data of biogas plants constructed by a company (without electricity generation)

	Digester 281 (B)	Digester 339 (DK)	Digester 81 (FRG)	Digester 283 (FRG)	Digester 150 (FRG)
a. Location of the plant	Baillonville	Graasten	Ahlen	Bad Waldsee	Bockenem
Type of waste	Poultry manure	Dairy cattle + pig manure	Bovine cattle manure	Poultry manure	Pig manure
Remarks					Not heated
Year of start-up	1981	1979	1979	1981	1981
Working volume (m^3)	21	2 x 190	50	120	280
b. Gas production rate ($m^3\ m^{-3}\ d^{-1}$)	0.9	0.8 - 0.9	1.5	0.80	1981-82 : 0.10 Thereafter : 0.20
c. Heating requirements (as % of b.)	30	33	25 - 35	-	0
d. Investment cost (ECU/m^3)	619	278	1 047	419	52.3
e. Yearly returns versus investment cost	0.12	0.10	0.05	0.07	1981-82 : 0.10 Thereafter : 0.20
f. Yearly operation cost versus investment cost	0.03	0.03	0.02	0.02	0.03
g. Pay-Back Period (years)	8	9	> 14	13	6
h. Internal Rate of Return (%)	11.7	7.8	- 5.9	0.72	17.9

Table I.4.4. Economic data of biogas plants constructed by a company (without electricity generation) (continuation)

	Digester 106 (FRG)	Digester 408 (FRG)	Digester 232 (I)	Digester 240 (NL)	Digester 345 (NL)
a. Location of the plant	Rhede	Wedemark	Vedelago	Klarenbeek	Witharen
Type of waste	Bovine cattle manure	Pig + bovine cattle manure + bedding	Pig manure	Pig manure	Dairy cattle + pig manure
Remarks				Operates only 6 months per year	
Year of start-up	1981	1981	1980	1979	1980
Working volume (m^3)	490	60	780	75	176
b. Gas production rate ($m^3\ m^{-3}\ d^{-1}$)	0.58	1.1	1	0.80	1980-81 : 0.31 Thereafter : 0.70
c. Heating requirements (as % of b.)	25	25	43	35	52
d. Investment cost (ECU/m^3)	359	453	339	342	221
e. Yearly returns versus investment cost	0.07	0.09	0.06	0.12	1980-81 : 0.05 Thereafter : 0.10
f. Yearly operation cost versus investment cost	0.02	0.02	0.07	0.02	0.02
g. Pay-Back Period (years)	12	10	infinite	8	11
h. Internal Rate of Return (%)	1.89	5.65	not significant	11.2	3.93

Table I.4.4. Economic data of biogas plants constructed by a company (without electricity generation) (end)

	Digester 386 (NL)	Digester 245 (UK)	Digester 279 (CH)	Digester 350 (CH)	Digester 57 (CH)
a. Location of the plant	Schijndel	Hitchin (Herts)	Vuiteboeuf	Bonnstetten	Neunkirch
Type of waste	Pig + poultry manure	Pig manure	Pig manure	Dairy cattle + pig manure	Dairy cattle manure
Year of start-up	1980	1978	1978	1978	1979
Working volume (m^3)	180	34	200	50	85
b. Gas production rate ($m^3\ m^{-3}\ d^{-1}$)	1980-81 : 0.21 Thereafter : 0.70	0.66 - 1.24	1.4	0.96	0.33
c. Heating requirements (as % of b.)	39	50	30	56	60
d. Investment cost (ECU/m^3)	262	177	377	462	82.7
e. Yearly returns versus investment cost	1980-81 : 0.03 Thereafter : 0.11	0.08	0.17	0.032	0.09
f. Yearly operation cost versus investment cost	0.02	0.04	0.035	0.023	0.036
g. Pay-Back Period (years)	10	>14	5	> 14	13
h. Internal Rate of Return (%)	4.75	- 1.22	21.62	- 19.37	0.86

Table I.4.5. Economic data of biogas plants constructed by a company (with electricity generation)

	Digester 6 (DK)	Digester 168 (DK)	Digester 190 (F)	Digester 175 (F)	Digester 151 (FRG)	Digester 17 (I)
a. Location of the plant	Haslev	Skals	Schoenen-Steinbach	Bernay-en-Champagne	Bredstedt	Fossano
Type of waste	Dairy cattle manure	Pig manure	Pig manure	Pig manure	Pig manure	Dairy cattle manure
Year of start-up	1979	1979	1981	1980	1979	1979/80
Working volume (m^3)	2 x 200	360	200	250	95	67
b. Gas production rate (m^3 m^{-3} d^{-1})	1979-81 : 0.39 Thereafter : 0.63	0.63	1.05	0.86	0.7	1.79
c. Heating requirements (as % of b.)	44	25	43	39	30	25
d. Investment cost (ECU/m^3)	302	189	644	515	578	880
e. Yearly returns versus investment cost	1979-81 : 0.05 Thereafter : 0.13	0.15	0.04	0.06	0.05	0.08
f. Yearly operation cost versus investment cost	0.01	0.02	0.02	0.02	0.03	0.03
g. Pay-Back Period (years)	8	6	>14	>14	>14	10
h. Internal Rate of Return (%)	8.98	19.4	- 9.25	- 0.86	- 12.41	8.87

Table I.4.5. Economic data of biogas plants constructed by a company (with electricity generation) (end)

	Digester 32 (NL)	Digester 31 (NL)	Digester 346 (NL)	Digester 346 (UK)	Digester 246 (UK)	Digester 278 (CH)
a. Location of the plant	Jutrijp	Hemelum	Bredevoort	Halrsworth	Newmachar	Uttewil
Type of waste	Dairy cattle manure	Dairy cattle manure	Dairy cattle + pig manure	Dairy cattle + pig manure	Pig manure	Pig manure
Year of start-up	1980	1980	1981	1980	1980	1980
Working volume (m^3)	200	150	460	350	680	80
b. Gas production ($m^3\ m^{-3}\ d^{-1}$)	1980-81 : 0.46 Thereafter : 0.8	1980-81 : 0.31 Thereafter : 0.74	1981-82 : 0.42 Thereafter : 0.82	0.57	0.51	0.75
c. Heating requirements (as % of b.)			25 - 30	60		35
d. Investment cost (ECU/m^3)	315	247	410	302	259	791
e. Yearly returns versus investment cost	1980-81 : 0.06 Thereafter : 0.10	1980-81 : 0.06 Thereafter : 0.14	1981-82 : 0.07 Thereafter : 0.14	0.09	0.02	0.007
f. Yearly operation cost versus investment cost	0.02	0.02	0.02	0.03	0.02	0.04
g. Pay-Back Period (years)	11	8	8	11	>14	>14
h. Internal Rate of Return (%)	2.99	9.22	10.51	4.43	Not significant	Not significant

Table I.4.7. Recalculated economic data of 3 biogas plants, assuming an improved gas production

	Digester 6	Digester 343	Digester 246
Gas production ($m^3\ m^{-3}\ d^{-1}$)	0.63	0.82	1
Yearly returns versus investment cost	0.13	0.14	0.14
Simple Pay-Back Period (years)	6	7	6
Internal Rate of Return (%)	16.6	15.0	17.4

Poor operation of the methane digester can be either a continuous or a temporary problem. For digester 246, a daily biogas production of only 0.5 m^3 per m^3 digester working volume is being obtained because the pig manure to be digested has a dry matter content of only 2 % and furthermore is not produced in sufficient quantity. If we suppose that for the same methane digester, a daily biogas production of 1 m^3 per m^3 digester working volume could be obtained, this digester would become profitable with a Simple Pay-Back Period of 6 years (see Table I.4.7).

In the case of some other methane digesters, the biogas production was low only during the first two years and thereafter it improved. These two years of poor operation are sufficient to lower the profitability of the plants. Supposing for example that digesters 6 and 343 had had their cruising biogas productivity from the beginning of the operation, their Simple Pay-Back Period would be reduced from 8 to 6-7 years with an Internal Rate of Return of 15 % (Table I.4.7).

It must be added that some plants were originally designed mainly as anaerobic wastewater treatment plants, and as such may well compete with aerobic treatment plants. In some cases, biogas use was initiated after some time and always improved profitability.

As regards the second important reason for non-profitability of biogas plants, i.e. too high an investment cost, nothing can be changed now. From this analysis it appears possible to construct biogas plants at a cost not exceeding 400-450 ECU per m^3 of digester working volume. Indeed, of the 27 biogas plants constructed by companies in 1980 or before, 15 were constructed for a price lower than 350 ECU/m^3. It can be concluded that an investment cost limit of 400-450 ECU/m^3 must be respected in order to ensure profitability.

CHAPTER V. ENERGETICS

In a biogas plant, part of the biogas produced is usually consumed within the process itself. Energy is required for heating the influent, for maintaining the methane digester at the required temperature, for mixing the digester content and, finally, for loading the methane digester. The survey did not produce a complete energy balance for any biogas plant. Only the fraction of biogas utilized as energy source for the different operations mentioned above is given for some biogas plants, and this fraction varies from plant to plant. The ranges elicited by the survey are reported in Table I.5.1.

Table I.5.1. Fractions of biogas used for the digester operation itself

Country	Share of total biogas production (as %)
- Belgium	25 - 40
- Denmark	25 - 44
- France	20 - 43
- F.R. Germany	18 - 38
- Italy	20 - 46
- The Netherlands	24 - 56
- United Kingdom	25 - 60
- Switzerland	30 - 60

As can be seen in Table I.5.1, the lower ranges of the biogas fraction used within the methane digestion operation are around 20-30 %. On the other hand, very high values for the higher range are reported for some countries. This happens when the methane digester is operating poorly. Two cases were encountered. In the first, there is not enough feedstock to load the methane digester at its optimal rate; consequently, mean detention times are too long and loading rates too low. A larger part of the small gas production is then required to maintain the digester at the required temperature. In the second case, the feedstock can be very dilute; as a result, a large fraction of the biogas is wasted for what is in fact the heating of large quantities of water.

Of the total amount of biogas consumed within the operation of the methane digester, it results from the survey that about 90 % of the energy demands serve to heat the influent and the methane digester. One exception was observed in Italy. There, about 35 % of the energy demand for the operation of the methane digester is devoted to mixing and loading. Indeed, in this large-scale Italian biogas plant, mixing occurs continuously.

Some exceptions were encountered during the survey. Agricultural methane digesters, operating under psychrophilic conditions (ambient temperature) need not be heated at all. And for a number of the industrial biogas plants, there are two major reasons for not using the biogas for the operation of the methane digester. First, some wastewaters are warm enough when they leave the production site; and secondly, many industries have waste steam in excess so that there is a preference to use that steam for heating the digester and to use the whole production of biogas for other purposes.

CHAPTER VI. INTEGRATION OF THE METHANE DIGESTER IN A BIOGAS PLANT

Besides the methane digester itself, three other aspects must be considered in a biogas plant. Upstream, the biomass resource must be stored and/or pretreated prior to biomethanation. Downstream, not only the biogas produced but also the liquid and/or solid effluents from the methane digester must be stored and/or post-treated prior to their further utilization.

VI.1. Storage and treatment of feedstocks

The feedstocks that are treated by biomethanation do not always enter the methane digester as they are produced : first they must either be stored or undergo pretreatment. This varies of course with the type of feedstock and with the type of biomethanation system. The various cases encountered in the survey are presented in the following paragraphs, first for agricultural wastes and then for industrial wastes.

VI.1.1. Agricultural wastes

a. Liquid or semi-solid wastes

The liquid or semi-solid agricultural feedstocks, i.e. the different types of manure, are in general stored during a very short time, from several hours to a few days. Nevertheless, exceptions were reported in The Netherlands where manure may be stored during a very long period of time, say up to 6 months. This was reported for four Dutch biogas plants where pig manure was stored for more than 2 months with the result that little biogas was produced upon biomethanation.

It is well known that pig manure has a high level of biodegradability, so that long storage periods lead to the escape of methane during storage. Moreover, pig houses are usually very warm and so are the storage tanks underneath the stables. These high temperatures favour the early methane digestion of pig manure in its storage tank. From experience, it can be recommended not to store pig manure prior to biomethanation, at least not for a long time. A maximum of 2 months storage is now applied in The Netherlands.

When treatment of the manures prior to biomethanation is considered, it appears from the survey that dilution of manures is one of the most frequent pretreatments. Although liquid manures are not treated at all, and this represents more than half the cases, dilution of more concentrated manures such as some dairy or beef cattle manures and poultry manure appears necessary in order to make them pumpable. Involuntary dilution by washing water and by drinker leaks appears to be a problem for pig manures, especially in large piggeries such as those in Italy. Pig manure is produced with a low content in dry matter, 5 % as total solids, and the addition of water results in very dilute manure with a concentration in dry matter falling to only 1 or 2 %. This dilute manure is not a good feedstock for completely mixed, once-through methane digesters. Larger working volumes of digesters are required without consequently higher gas production. In Italy, prethickening of the pig manure is therefore sometimes practised.

Among the other pretreatments mentioned, grinding of manure is often practised. A tearing pump placed inside the storage tank serves to cut pieces of straw or roughages that may be contained in the manure. A less frequently practised pretreatment of the manure is biological warming by aeration. This is a practice encountered in Switzerland where it is effected in the Roxistall system. The methane digester constructed under the stable has an integrated biological aerated tank.

Special attention must also be given to the development of the anaerobic filter biomethanation system which requires settlement by gravity and possibly dilution of the pig manure prior to biomethanation.

b. Solid wastes

Solid agricultural wastes generally undergo a period of storage of 1 to 3 months. Indeed, manure with bedding is often disposed of in heaps after its production. This heaping leads to a natural aerobic fermentation which raises the temperature of the manure. Manure with bedding which is digested in batch systems is not treated prior to biomethanation, but when it is biomethanized in continuous, once-through, completely mixed or plug-flow systems, the manure and the straw are ground and diluted. Addition of lime has been reported for goat manures with bedding which are more acid.

In pilot experiments on the biomethanation of energy crop residues, two-step biomethanation systems can be considered. The first step may be either a physical step of grinding and maceration or a biological step of fermentative liquefaction or a combination of both. The second step is then typically the one-step, once-through, continuous biomethanation system. The first step can also be a percolation step, again either physical or biological or a combination of both. For the more liquid effluent of percolation, biomethanation systems with accumulation of active biomass such as the upflow system without carrier are considered.

VI.1.2. Industrial wastewaters and wastes

No storage of industrial wastes or wastewaters has been reported. The pretreatment most applied to wastewaters is neutralization with lime. Neutralization is often mentioned for distillery and pectinery wastewaters but seldom for the wastewaters of sugar refineries and potato factories.

In some cases, cooling of the wastewaters prior to their biomethanation is necessary. There is one report of preliminary denitrification of pectinery wastewater. Finally, grinding or screening of the solid part of wastewaters is reported twice.

Pilot-scale experiments in completely mixed, once-through, continuous biomethanation systems on slaughter-house wastes require dilution and grinding of the feedstock. Pilot-scale experiments in the batch system on the same solid industrial wastes may be preceded by an aerobic fermentation step.

VI.2. Uses of the biogas

The different ways of storing and purifying the biogas prior to its utilization have been discussed in Chapter II.
As a general rule, part of the biogas is used for the methane digestion process itself, as explained in the preceding Chapter. The remainder is used either for professional or domestic purposes.

a. Uses on farms

The proportions of biogas used either for domestic or professional purposes depend essentially on the type of farming. It appears from the survey that professional uses of biogas are more important than domestic uses only in the case of farming with continuous specific energy needs. Where no specific needs exist, the biogas is preferably used for domestic purposes.

The specific energy needs of the farm may also justify the installation of an engine. Indeed, it appears from the survey that about one third of the biogas plants have at least one engine for electricity generation.

Domestic uses consist essentially in heating and the production of hot water. Cooking with biogas or the production of domestic electricity from biogas are more rarely practised.

The professional uses of biogas are clearly specific to the type of farming. They can be heating of housing for calves or young pigs, dairy or other electric power, hot water for cleaning, feed cooking, fodder drying, milk pasteurization, etc.

Biogas can also be sold, either as gas or after conversion into electricity. However, conditions for the sale of biogas by individuals are unfavourable in most countries (see Part II, Chapter III). This explains why the sale of energy produced from biogas plants is not a common practice. Very few owners of biogas plants are reported selling part of their gas or electricity production : 2 in Italy, 2 in Denmark, 2 in The Netherlands and 1 in the Federal Republic of Germany. With the exception of one biogas plant owner in Italy who sells 57 % of his total methane production as gas, the others are selling less than 7 % of the total methane production as electricity.

b. Uses in industry

As seen in the preceding Chapter, most of the industries use all the biogas produced for professional purposes. Biogas is used mostly for heating or steam production. Pulp drying using biogas is characteristic of sugar refineries.

It should be noted that, at the time of the survey, not all the industries were using their production of biogas as an energy source. Sometimes the biogas was wasted and burned in torches. Nevertheless, most of these industries foresaw useful applications for the biogas in the future.

VI.3. Treatment and uses of the digested effluents

Further treatment and possible uses of liquid and solid effluents after methane digestion differ according to whether they originate on farms or in industry.

a. Agricultural effluents

Agricultural wastes are often used without treatment after biomethanation. Nevertheless, in a few countries such as Italy the available land is insufficient to receive the effluents of intensive breeding as in large piggeries. Since the laws on limits of pollution by agricultural wastes are enforced in Italy, further treatments are necessary in order that the wastes may be discharged to watercourses and/or in order to reduce the quantities of wastes applied to land. Some of the treatments are practised with a view to either further depollution (such as aeration with or without activated sludge) or better fertilizing value (*e.g.* composting). Table I.6.1 presents a summary of the different treatments already practised or foreseen according to the type of manure.

Digested agricultural effluents are often used as soil conditioners and/or fertilizers. As can be seen in Table I.6.1, and often after some particular treatment such as aeration, clarification and/or filtration, the solid fraction of the digested wastes only is used on soil, whereas the liquid part is discharged into watercourses.

Fresh effluents of methane digesters are hardly ever sold. Two cases in Italy are an exception. As a general rule, the sales of digested effluents took place only after composting.

b. Industrial effluents

In contrast to the agricultural digested effluents, post-treatment of industrial digested wastes is a common practice. Indeed, the latter are at least disposed of in lagoons, and in some cases are even treated aerobically. Particular post-treatments typical of certain industrial wastes or wastewaters are presented in Table I.6.2.

As to their final disposal, post-treated digested industrial wastewaters are mainly discharged to sewers or to municipal wastewater treatment plants. In some industries such as sugar refineries, the liquid effluents are recycled into the factory. The excess sludges remaining after the post-treatment of the digested wastewaters are sometimes used as soil conditioner. Solid industrial digested wastes serve also for the latter purpose but after composting.

Table I.6.1. Treatments and uses of digested agricultural wastes

Type of digested agricultural waste	Treatment	Uses and disposal
- Bovine cattle manure	- Settled	Used on soil
- Pig manure	- Disposed of in lagoon - Disposed of in drying beds	Used on soil
	- Aerated with activated sludge - Aerated and clarified	Discharged to watercourse
	- Settled, and solids possibly composted thereafter - Clarified and treated(1) chemically; solids pressed in filter plate press and liquids disposed of in lagoon - Filtrated, treated by activated(1) sludge and settled; solids composted	Solids used on soil and liquids discharged to watercourse
- Poultry manure	- Settled, flocculated and filtrated	Used on soil
- Bovine cattle manure with bedding	- Dried	Sold as soil conditioner

(1) Foreseen

VI.4. Recommendations for the integration of a biogas plant

Whether a biogas plant is to be integrated into a farm or an industry, some lessons learnt from the survey can be drawn on to guide the construction of future plants.

The choice of an appropriate location for the methane digester is very important for a good integration of the biogas plant. The methane digester must be located as near as possible to the feedstock storage tank, for two reasons : first, the feedstock can then be biomethanized while it is still fresh without even being completely cooled; secondly, excessive loading losses can be avoided in this way. The methane digester must also be located at a lower level than the storage tank, again to avoid loading losses. With this requirement in view, the methane digester may even be partly buried. Finally, the methane digester must be located so as to ensure protection against the wind and the cold, for example behind a line of trees or a wall. In winter this can be very important to avoid heat losses and the risk of frozen pipes.

Table I.6.2. Treatments and uses of digested industrial wastes

Type of digested industrial waste	Type of treatment	Uses and disposal
- Sugary wastewater	Not treated or disposed of in lagoons	Recycled into the factory
- Distillery wastewater	Biologically treated or flocculated	Water discharged to sewer, and sludge used as soil conditioner
- Starch factory wastewater	Freshed-up or settled and disposed of in lagoons	Water discharged to sewer, and sludge used as soil conditioner
- Pectin industry wastewater	Settled and treated by aerobic activated sludge	Water discharged to municipal wastewater treatment plant, excess sludge used on soil
- Slaughter-house waste (after batch digestion)	Composted	Used as soil conditioner
- Potato processing wastewater	Not treated or aerated, nitrified and clarified	Discharged to sewer or to municipal wastewater treatment plant
- Vegetable canning wastewater	Settled and disposed of in lagoon	Discharged to sewer

Successful integration of a biogas plant presupposes also an effective utilization of the biogas produced. So far as possible, the biogas should be used within the farm or industry. This assumes of course that there is a regular and more or less constant need for energy, as in industry. However, farmers who have a disposable surplus of biogas can make local arrangements with a nearby industry or any other user of energy. It should be remarked here that if there is no regular or important need for electricity within the farm there is no valid reason to install an engine that converts the biogas into electricity. A lot of money and worries can then be avoided.

A general remark is that the operation of a biogas plant must be as simple as possible for the user. This does not imply a complete automatization of the plant, but loading and mixing of the methane digester should be easy. This is especially true for the farmer who has not much time to spare.

Finally, selecting a biomethanation system well adapted to the feedstock to be treated is also extremely important for effective integration. There exist in 1983 a number of improved technologies which offer improved performances. The best example is certainly the biomethanation of pig manure.

It appears from the survey that it is often the classical completely mixed system which is applied, even with very dilute pig manure. For such feedstocks, the anaerobic filter is certainly better indicated : for the same quantity of wastes, and the same yield of biogas, the digester working volume can then be reduced at least fivefold.

Allerhop (F.R. Germany) 1950. Biogas compressor. Picture from Def Mitteilungen, Heft 9, 1951. Published in Verden.

Faenza (Italy) 1981. Compressor for gas mixing of a completely mixed continuous methane digester on distillery wastewater.

I-FAENZA - DIGESTER 492

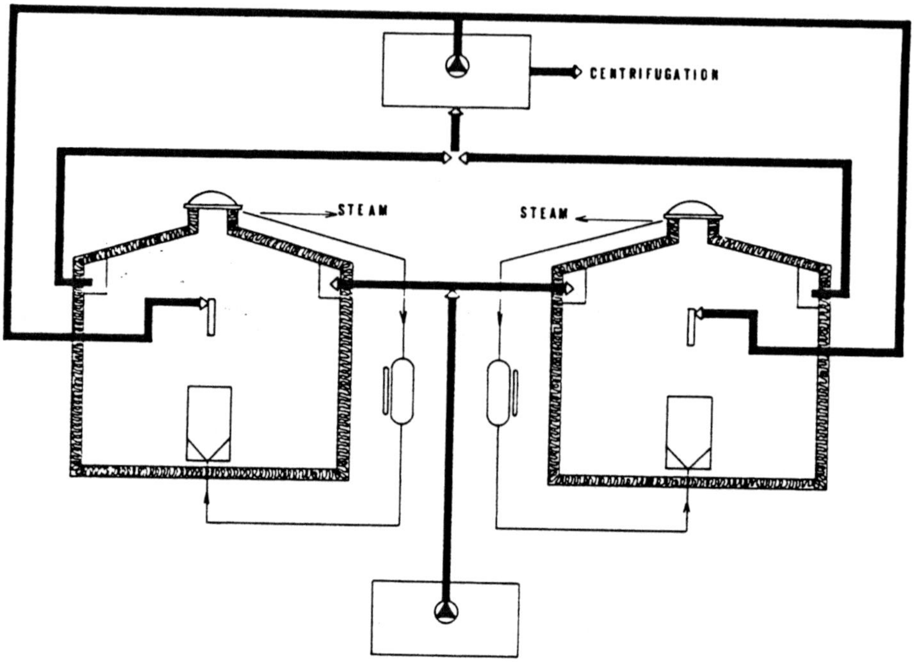

ADDRESS : NERI, Distillery - FAENZA (Ravenna) Italy
START-UP : April 1981
DESIGN OF THE DIGESTER :

- System : completely mixed, continuous loading
- Volume : total = 2 x 7 000 m^3, working volume = 2 x 6 000 m^3
- Material : steel
- Insulation : expanded polyurethane
- Mixing system : gas lift
- Heating system : not necessary since T of feedstock about 40 °C

COST : Not available

FEEDSTOCK : Distillery wastewaters (dreg, wine, fruits)
TREATMENT : Cooling to 40 °C, neutralization...
RUNNING CONDITIONS :
B_V = 2.7 kg VS m^{-3} d^{-1}
T = 35 °C
θ_H = 10 d
θ_S = 10-25 d

GAS :
- Storage : gasometer (500 m^3)
- Production : rate : 1.5 m^3 d^{-1} m^{-3}
  : yield : 0.55 m^3 kg^{-1} VS_o
- Composition : 65 % CH_4
- Depollution : 55 % VS reduction
- Uses : 100 % professional use (steam)

MAJOR PROBLEMS :
Settling of digested mixed liquor resolved by centrifugation

CHAPTER VII. DECISION-MAKING IN DIGESTER DESIGN ACCORDING TO FEEDSTOCK CHARACTERISTICS

VII.1. Introduction

Recent progress in the microbiology and biochemistry of the methanogenic process (reviewed by Nyns, 1981) has made it possible to construct a decision model for the choice of the methane digestion system most appropriate to the characteristics of the substrate biomass to be used as feedstock. The characteristics of the substrate biomass to be taken into consideration are, first, the physical state of the substrate biomass and, secondly, its fermentation profile.

A substrate biomass may be either semi-solid or solid, or semi-liquid or liquid. A semi-solid or solid substrate biomass is basically not pumpable. A semi-liquid substrate biomass or slurry is basically pumpable, although it often contains substantial amounts of suspended solids. Its concentration in dry matter exceeds 4 % (wt wt^{-1}). A liquid substrate biomass or a wastewater is pumpable. It contains basically few if any settleable suspended solids. It may contain, however, substantial amounts of unsettleable, among others colloidal, suspended solids. Its concentration in dry matter usually does not exceed 4 % (wt wt^{-1}).

By fermentation profile, the following concept is meant. When an anaerobic complex microbiological ecosystem is allowed to evolve spontaneously under septic conditions, the course of fermentation, as elicited by the pattern of metabolites as a function of time, appears reproducible provided physico-chemical parameters are identical, irrespective of the biology of the ecosystem. This suggests a control of the ecosystem, in a unique way, more by the physico-chemical parameters than by its biology (Legros _et al._, 1983).

The major fermentation profiles encountered hitherto under anaerobic conditions are the following :

a. acid fermentation profiles :
   - prevalence of lactic acid (pH below or equal to 4.5)
   - prevalence of volatile fatty acids with 2 to 5 carbon atoms (pH below or equal to 6.5)

b. neutral fermentation profile (pH around 7) :
   - prevalence of lactic acid
   - prevalence of volatile fatty acids with 2 to 5 carbon atoms
   - prevalence of ethanol
   - prevalence of methane.

The biomethanation systems subject to decision-making are the following :

a. batch systems :
   - with one unique loading or simple batch system
   - with multiple loading or fed batch system

b. continuous systems :
- once-through, completely mixed system
- once-through, plug-flow system
- with active biomass recycle
- with active biomass entrapment, with or without immobilization
- one or two-step systems.

Knowing the fermentation profile of the candidate substrate biomass, which can easily be determined experimentally (Legros et al., 1983) and its physical state, the grid constructed in the following paragaphs gives directly the most appropriate biomethanation system.

VII.2. Grid of the decision-making model

VII.2.1. Case of semi-solid or solid substrate biomasses

The criterion of choice is selected from the three following possibilities :

a. It is possible and economically feasible to dilute the solid or semi-solid substrate biomass to a final dry matter content comprised between 4 and 10 % (wt wt^{-1}) and to homogenize its content in suspended matter so as to reduce the particle size below 6 mm. This case can then be assimilated to that of the semi-liquid substrate biomasses, considered below under VII.2.2.

b. The solid or semi-solid substrate biomass exhibits a natural tendency towards liquefaction. This is the case of numerous non-lignified plants. Liquefaction may be a physical process, e.g. the homogeneization of water hyacinths which usually have a water content equal to 80 % of their weight. Liquefaction may also be a biological process, e.g. the spontaneous fermentation of algae. In these cases, liquefaction can be allowed to proceed and the resulting substrate biomass treated as for the semi-liquid substrate biomasses, considered below under VII.2.2.

c. In all other cases, the choice is based on the fermentation profile, F, which can be either acid, F_A, or neutral, F_N.

a. Case of acid fermentation profiles of semi-solid or solid substrate biomasses

Whenever the fermentation profile, F, of a semi-solid or solid substrate biomass is acid, F_A, and this is the case of numerous plants or plant residues, there exist three possibilities of choice.

1. The mixing of the solid or semi-solid substrate biomass with an acid fermentation profile, $S_{solid\ FA}$, with a solid or semi-solid substrate biomass with neutral fermentation profile, $S_{solid\ FN}$, so as to obtain a mixture which globally exhibits a neutral fermentation profile, $S_{solid\ FN}$. In this case, the most appropriate methane digestion systems are the simple or fed-batch systems :

$$S_{solid\ FA} + S_{solid\ FN} \longrightarrow \Sigma S_{solid\ FN} \Longrightarrow \text{Batch} \qquad (VII.1.)$$

2. The alternate loading of the methane digester with the solid or semi-solid substrate biomass with an acid fermentation profile, $S_{solid\ FA}$, and with a solid or semi-solid biomass substrate with a neutral fermentation profile, $S_{solid\ FN}$. In this case, the most appropriate methane digestion system is the fed-batch system :

$$(S_{solid\ FN} + S_{solid\ FA} + S_{solid\ FA} + \ldots)_{n\ times} \Longrightarrow \text{Fed batch} \quad (VII.2.)$$

3. A two-step biomethanation process. In this case, there exist two possibilities of choice :

- Either liquefy or dilute the solid or semi-solid substrate biomass with an acid fermentation profile as described under VII.2.1. a and b, and treat as in the case of semi-liquid substrate biomasses to be considered under VII.2.2.

- Or design the first step of the biomethanation two-step process according to the principle of percolation. By this process, a solid or semi-solid substrate biomass with an acid fermentation profile, $S_{solid\ FA}$, is transformed, on the one hand into a liquid product with a neutral fermentation profile, $P_{liquid\ FN}$ to be considered either as a semi-liquid or as a liquid biomass substrate (see below, VII.2.2 and 3), and, on the other hand, into a semi-solid product with a neutral fermentation profile, $P_{solid\ FN}$, to be treated as proposed below under VII.2.1.b.

$$S_{solid\ FA} \xrightarrow[\text{Percolation}]{} P_{liquid\ FN} + P_{solid\ FN} \quad (VII.3.)$$

The two former possibilities of choice (1 and 2) have often and on been mentioned in the literature but still need clear-cut demonstration. The third possibility of choice has been well documented by Rijkens (1981). Percolation is the most appropriate, if not the only choice whenever the fermentation profile is strongly acid, because percolation exhausts the acidifying tendency of the substrate biomass.

b. Case of neutral fermentation profiles of semi-solid or solid substrate biomasses

Whenever the fermentation profile of a solid or semi-solid substrate biomass is neutral, S_{FN}, and this is the case for numerous animal wastes, two biomethanation systems enter into consideration but no particular criterion of choice exists between them. These are the simple and fed batch systems :

$$S_{FN} \Longrightarrow \text{Simple or fed batch} \quad (VII.4.)$$

VII.2.2. Case of semi-liquid substrate biomasses

Semi-liquid substrate biomasses include slurries in general. The criterion of choice of an appropriate biomethanation system is the fermentation profile, F.

a. Case of acid fermentation profiles of semi-liquid substrate biomasses

Whenever the fermentation profile, F, of a semi-liquid substrate biomass is acid, F_A, and this is the case of various non-lignified plant residues, three biomethanation systems enter into consideration.

1. The alternate loading of a one-step, once-through, completely mixed continuous methane digester with a semi-liquid substrate biomass with an acid fermentation profile, $S_{semi-liquid\ FA}$ and with a semi-liquid substrate biomass with a neutral fermentation profile $S_{semi-liquid\ FN}$:

$$(S_{semi-liquid\ FA} + S_{semi-liquid\ FN})_{n\ times} \Longrightarrow \text{One-step, completely mixed} \quad (VII.5.)$$

2. Methane digestion in two steps, each once-through, completely mixed and continuous. In the first step of acidifying fermentation, the acid fermentation profile of the semi-liquid substrate biomass is eroded, allowing methanogenesis to take place in the second step in a reliable way :

$$S_{semi-liquid\ FA} \Longrightarrow \text{Two step, completely mixed} \quad (VII.6.)$$

3. Whenever the acid fermentation profile is strongly marked, two cases can be distinguished :

- Either the semi-liquid substrate biomass has a physical state which permits percolation. This is treated as case 3 under VII.2.1.

- Or the semi-liquid substrate biomass does not exhibit such a physical state, in which case it must be diluted and treated as the corresponding liquid substrate biomass under VII.2.3. The typical case is the high-strength wastewater from olive oil mills.

b. Case of neutral fermentation profiles of semi-liquid substrate biomasses

Whenever the fermentation profile of a semi-liquid substrate biomass is neutral, $S_{semi-liquid\ FN}$, and this is the case of most animal wastes, manures and stomach contents (paunch manure), the most appropriate biomethanation systems are the one-step, once-through, completely mixed or plug-flow continuous systems :

$$S_{semi-liquid\ FN} \Longrightarrow \text{One-step, completely mixed or plug-flow} \quad (VII.7.)$$

VII.2.3. Case of liquid substrate biomasses

The criterion of choice for a biomethanation system most appropriate to a given substrate biomass is once again the fermentation profile, F.

a. Case of acid fermentation profiles of liquid substrate biomasses

Whenever the fermentation profile of a substrate biomass is acid, $S_{liquid-FA}$, the most appropriate biomethanation systems are the two-step systems :

$$S_{liquid\ FA} \Longrightarrow \text{One-step, completely mixed + one-step with accumulation of active biomass} \quad (VII.8.)$$

The first step of acidifying fermentation can be simple, micro-aerophilic or anaerobic. It may take place at ambient temperature and last a short time. The second step of methanogenesis should be one of the biomethanation systems with accumulation of active biomass, as shown in Table 0.1.

However, whenever the acid fermentation profile is strongly marked, it appears necessary to choose a single-step system with accumulation of active biomass for the global biomethanation process and to dilute the liquid substrate biomass.

b. Case of neutral fermentation profiles of liquid substrate biomasses

Whenever the fermentation profile of a liquid substrate biomass is neutral, $S_{\text{liquid FN}}$, single-step biomethanation systems with accumulation of active biomass suffice :

$$S_{\text{liquid FN}} \Longrightarrow \text{One-step with accumulation of active biomass} \qquad \text{(VII.9.)}$$

VII.3. Side criteria of choice

Side criteria of choice can influence the final decision on a biomethanation system most appropriate to a given substrate biomass. They include the utilization of end-products and the side effects of the biomethanation process. They take into account the need for intensive energy production or on the contrary a potential excess production of energy. These side criteria of choice are :

- depollution,
- hygienization or killing of pathogens,
- production of semi-liquid or almost dry fertilizer,
- production of semi-liquid or almost dry animal feed.

Finally, integration of the biogas plant in its surrounding must be taken into account. The decision will depend upon whether the biogas plant is to be integrated into a family residence, a farm, or an industry.

PART II. STATUS OF BIOMETHANATION

CHAPTER I. STATUS OF BIOGAS PLANTS IN THE EUROPEAN COMMUNITY AND IN SWITZERLAND

The survey has permitted the identification of 546 biogas plants. Information has been gathered on the type of waste treated, the location of the plants and their constructors, the total number and its evolution with time. This information is scrutinized in the present chapter. For each biogas plant, detailed information is available in Appendix III which lists all biogas plants in the European Community and in Switzerland.

The survey deals mainly with biomethanation processes of slurries, semi-solid and solid wastes with the production of energy as primary aim. Landfills are included. Attention is paid also to processes devoted to wastewaters, even though they have pollution control as primary target.

I.1. The number of biogas plants

A total of 546 biogas plants had been identified by the end of 1982 in the different Member States of the European Community and in Switzerland (Table II.1.1). Four hundred and twenty biogas plants (among which 378 full-scale plants and 42 pilot-scale plants) are treating agricultural wastes, whereas 89 biogas plants (among which 69 full-scale plants and 20 pilot-scale plants) are treating industrial wastes or wastewaters. As regards the biomethanation of energy crops, there seems to be only one pilot plant in the south of Italy, at Lamezia. Finally, the abstraction of biogas from landfills is at present being carried out at 23 full-scale and 13 test sites.

Whereas the number of biogas plants working on agricultural wastes is the largest with 77 % of the total number, it has been calculated that the overall working volume of methane digesters on farms is smaller than that of those working on industrial wastes. The survey indicated a total of 95 000 m^3 of methane digester working volume for agricultural wastes as compared with 174 000 m^3 for industrial wastes. For landfills, it was not possible to estimate the total working volume.

I.2. Geographical distribution of biogas plants

The geographical distribution of the biogas plants within the different countries can also be seen in Table II.1.1 and is depicted in Fig. II.1.1. Methane digesters on farms are found in all the countries. Almost the same can be observed for biogas plants on industrial wastes with the exception of Switzerland where there seem to be as yet few if any biogas plants in industry.

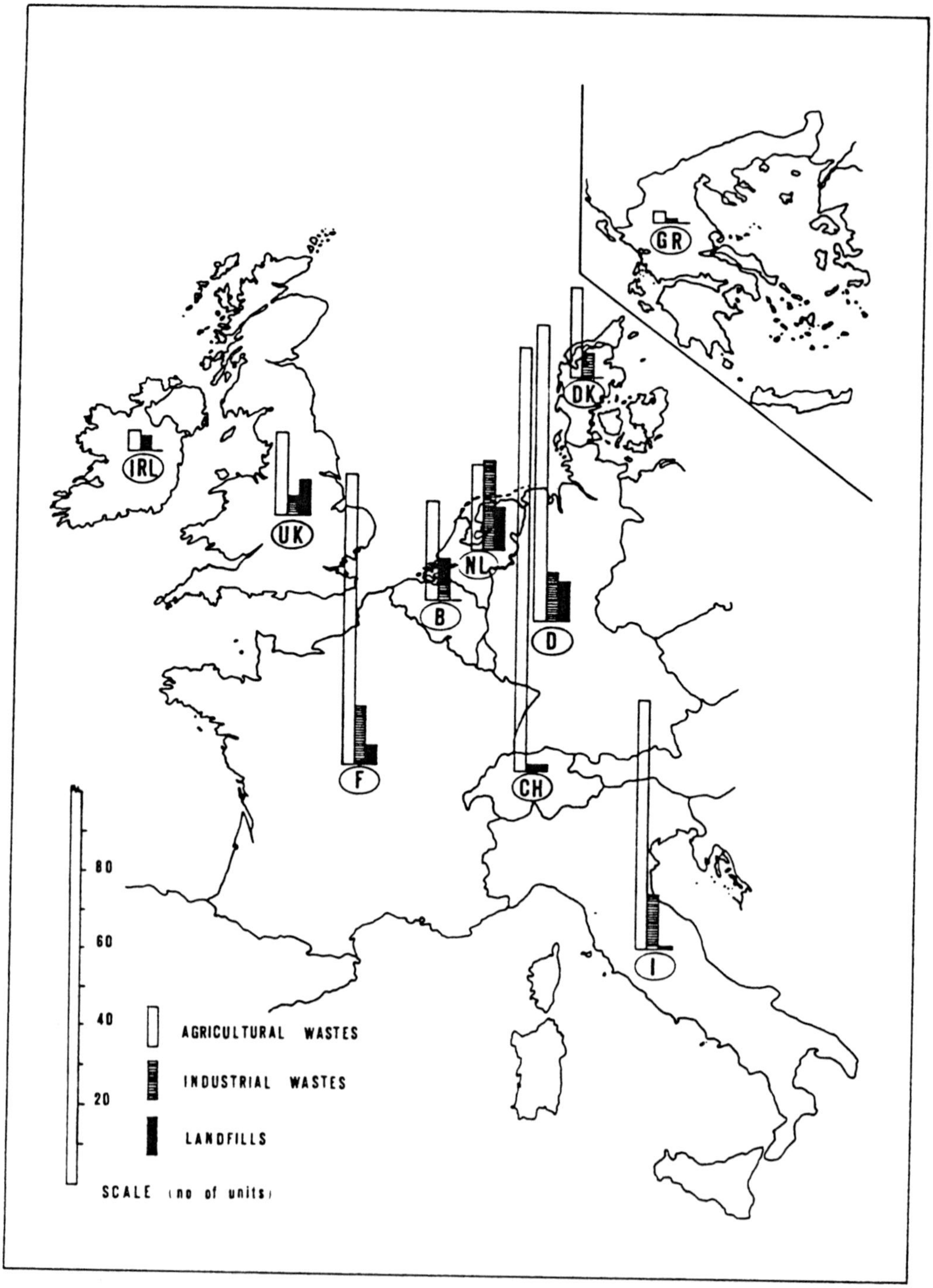

Fig. II.1.1. Distribution of biogas plants in the European Community and Switzerland

Table II.1.1. Distribution of biogas plants in the European Community and in Switzerland according to the type of waste treated

Country	No. of plants per type of waste				
	Agricultural wastes	Energy crops	Domestic refuse	Industrial wastes	Total
Belgium	25			10	35
Denmark	23			6	29
France	74		5	15	94
F.R.Germany	75		10	12	97
Greece	4			1	5
Ireland	5			4	9
Italy	63	1	1	13	78
The Netherlands	22		11	23	56
United Kingdom	21		9	5	35
Switzerland	108				108
Total	420	1	36	89	546

Table II.1.2. Number of biogas plants per type of agricultural waste

Country	Liquid or semi-solid	Solid	Mixed	Total
Belgium	21	1	3	25
Denmark	18		5	23
France	32	32	8	72
F.R. Germany	57	10	7	74
Greece	2		2	4
Ireland	5			5
Italy	58		5	63
The Netherlands	14	1	7	22
United Kingdom	11	1	9	21
Switzerland	73		31	104
Total	291	45	77	413

The exploited landfills appear to be less well developed, although five countries of the European Community are already concerned by this third category of biogas plants. In Switzerland there might be two landfills in exploitation. Since no information was available on the latter, they were not included in Table II.1.1 and are not summed in the total number of biogas plants.

The scale of the biogas plants, either full-scale or pilot-scale, is indicated in Appendix II.B., Table 2.1.

I.3. Distribution of biogas plants among categories of wastes

Within the three categories of wastes treatable by biomethanation, several particular types of wastes can be considered. The agricultural wastes, for instance, can be classified according to their physical state into three major categories : liquid, semi-solid, or solid. The classification can further take into account whether they are biomethanized alone or mixed with one or more other wastes. In this way, 5 different types of liquid or semi-solid manures, 11 different types of solid agricultural wastes, and 21 types of combinations of different agricultural wastes, either liquid or solid, have been distinguished.

Among the liquid or semi-solid wastes, cattle and pig manure are as yet the most common types of wastes to be biomethanized. The solid agricultural wastes are mainly manure with bedding. The mixed agricultural wastes are often mixtures of two or three different manures, although manure with bedding can be associated with manure prior to biomethanation. In some cases manure can also be biomethanized together with industrial wastewaters.

Table II.1.2 gives the number of biogas plants per major category of agricultural waste. It will be seen that most of the biogas plants are treating liquid or semi-solid agricultural wastes. The biogas plants treating solid agricultural wastes are mainly located in France and in the Federal Republic of Germany. Finally, the 77 plants on mixed agricultural wastes are spread widely in all the countries.

Tables 2.2, 2.3 and 2.4 in Appendix II.B. detail the number of biogas plants for each of the 37 types of agricultural wastes and for each country.

At present there are 21 different types of industrial wastewaters and two different types of industrial wastes which serve as substrate biomass or load for biomethanation in industry. Both the wastewaters and the more solid wastes are mainly of agro-food origin, although three of the industrial wastewaters originate from the leather, wooden plates and paper industries. Table 2.5 in Appendix II.B. details the number of methane digesters for each type of industrial waste. It can be noted here that the wastewaters of sugar refineries, distilleries and potato-processing industries are the most common types of industrial wastes submitted to biomethanation.

I.4. Survey of biogas plants per country

For each country of the European Community and for Switzerland, the full-scale biogas plants have been located on a map. Whenever possible, a distinction has been made between full-scale plants in project or under construction and full-scale plants already on stream before, or starting up during 1982. Each biogas plant listed in Appendix III can be located in a square defined by a combination of one letter and one figure.

a. Belgium

Fig. II.1.2 shows that the methane digesters in Belgium are spread all over the country. Those located in the northern part of the country are mostly treating pig manure, whereas those in the south are mostly treating bovine cattle manure or sugary wastewaters. It can also be observed that many plants on farms are still under construction. The reason for this is that a financial support programme for biogas plants on farms was initiated during 1982 and is continuing in 1983.

b. Denmark

In Denmark there is also a large spread of the biogas plants over the whole country, as shown in Fig. II.1.3. Some plants on farms, started up at the end of the seventies are now shut down because of several technical problems and the absence of monitoring. It can also be observed that, with the exception of one biogas plant in industry in operation since 1969, the interest in biomethanation of industrial wastewaters is of recent origin.

c. France

The biogas plants located in France are presented in Fig. II.1.4. The first methane digesters on farms were erected in the areas of Brittany, Loire, Rhône, Alps and Alsace, all regions where intensive breeding is practised. The more recent biogas plants are widespread over the whole country. Biogas plants in industry are mostly located in the north of France where sugar refineries and distilleries are concentrated.

d. Federal Republic of Germany

A synopsis of the regional distribution of the methane digesters in the Federal Republic of Germany is presented in Fig. II.1.5. In the north the biogas plants are widely spread, whereas in the south an important concentration of biogas plants is located around the two research centres, located at Stuttgart (Universität Hohenheim) and Munich (Landtechnik Weihenstephan Institute) where the first German biogas plants were constructed.

e. Greece

It will be seen in Fig. II.1.6 that there are in Greece three projected biogas plants on farms and one biogas plant in operation in industry. One of the agricultural digester projects and the industrial plant are located near the research centres of the Technical University and the Agricultural College of Athens. The two remaining plants on farms are located near the

Research Centre of the Aristotle University of Thessaloniki.

f. Ireland

Fig. II.1.9 locating the biogas plants in the United Kingdom and in Ireland shows the three Irish full-scale biogas plants under construction.

g. Italy

Fig. II.1.7 gives the regional distribution of the biogas plants in Italy. Most of the agricultural biogas plants are located in the north where there are "industrial breeding units" of pigs. Nevertheless, for the south, there is a large project of about 30 biogas plants which will be constructed in the coming years.

h. The Netherlands

The methane digesters of The Netherlands are presented in Fig. II.1.8. Since pig farms are concentrated in the east and the south of the country, most piggery waste digesters are found in these regions. Dairy manure digesters are distributed all over the country, but most are erected in the north where dairy cattle breeding is the predominant branch of agriculture.

i. United Kingdom

As can be observed in Fig. II.1.9, most of the biogas plants, not only agricultural and industrial but also landfills, are concentrated in the southern part of the United Kingdom where most of the research centres are also located.

j. Switzerland

Fig. II.1.10 presents all Swiss biogas plants existing in February 1982. It can be seen that biogas has made a particular impact in the northern and central part of Switzerland. This is explained by the fact that most of the large farms are concentrated in that area. The two industrial plants and the landfills which were not included in the listing presented in Appendix III are nevertheless located on the map.

I.5. Evolution of the number of biogas plants with time

It has been seen in the preliminary chapter on biomethanation and biogas plants that some biogas plants already existed in France and Germany in the fifties. Nevertheless, most of them are now shut down. Two French methane digesters, started up respectively in 1952 and in 1956, are still in operation in 1983. But one has to wait until the end of the sixties and the beginning of the seventies to see the real start of new agricultural methane digesters. Between 1956 and 1977, only 12 plants were constructed, and it is only since 1978-1979 that the number of agricultural methane

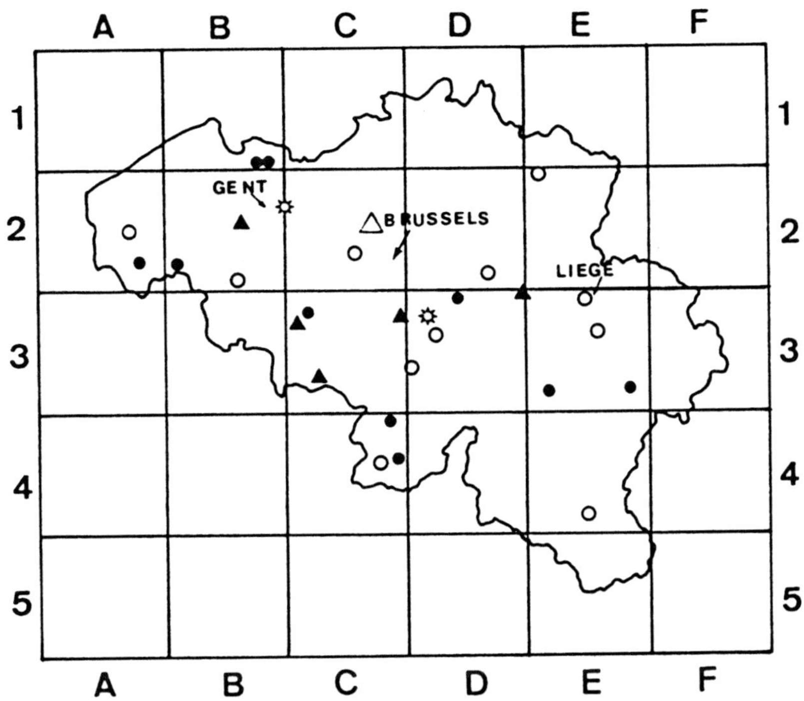

Fig. II.1.2. <u>Location of biogas plants in Belgium</u>

● : Plants on stream on farm; ▲ : Plants on stream in industry; ○ , △ : Plants in project or under construction; ✲ : Research centres.

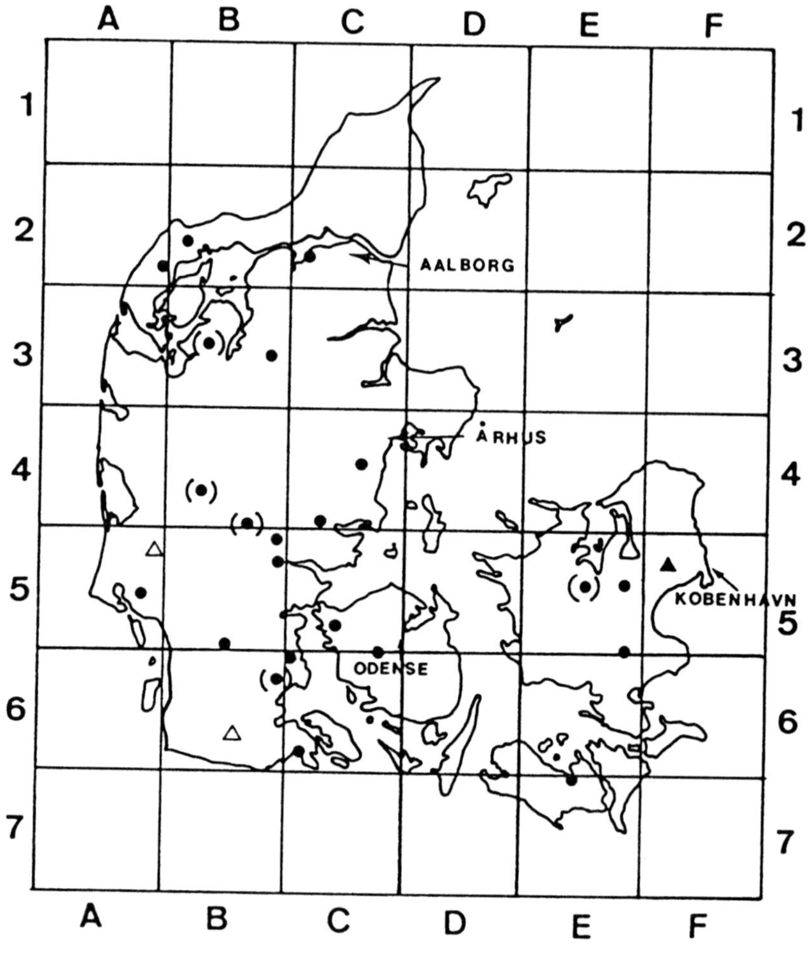

Fig. II.1.3. Location of biogas plants in Denmark

●: Plants on stream on farm; (●) : Plants on farm shut down; ▲ : Plant on stream in industry; △ : Plants under construction in industry.

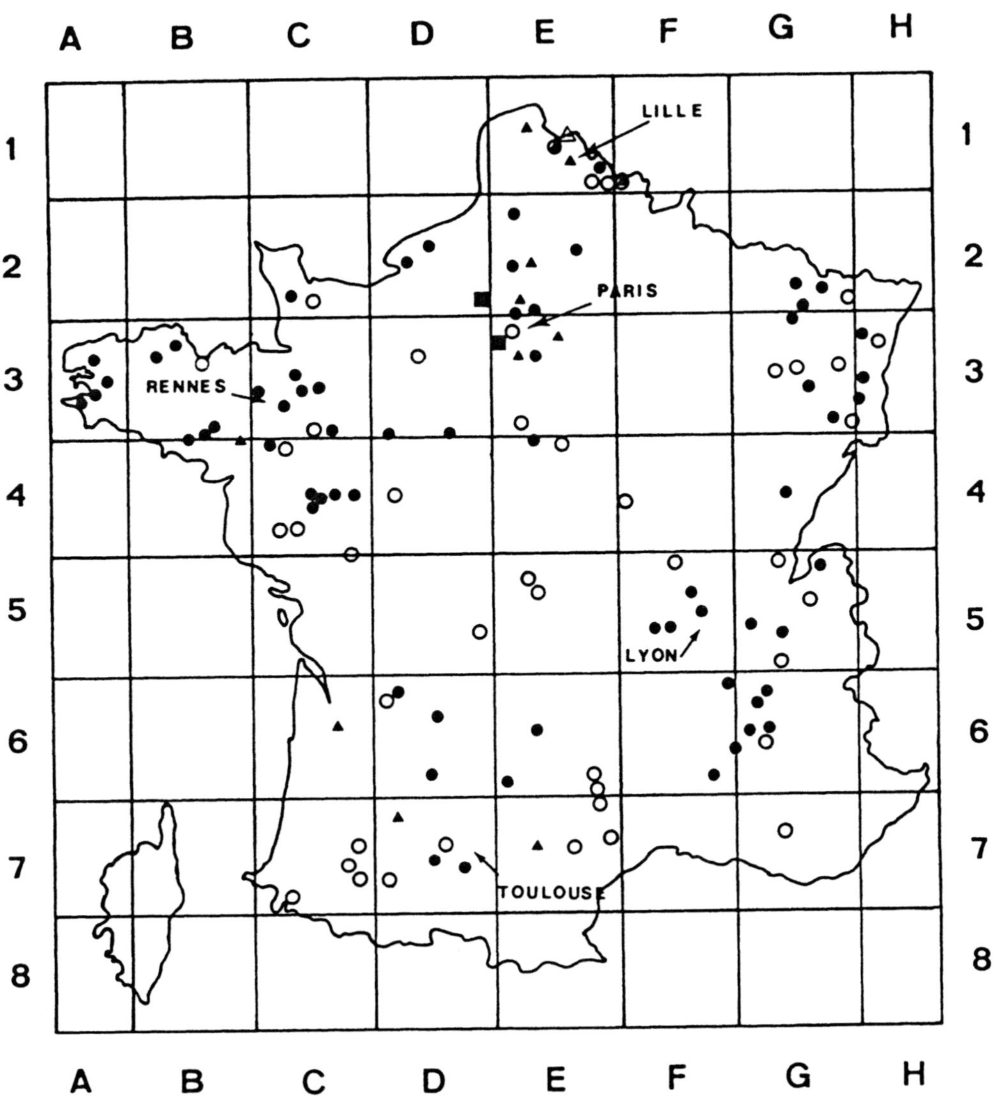

Fig. II.1.4. Location of biogas plants in France

●: Plants on stream on farm; ▲ : Plants on stream in industry; ■ : Landfills; ○ , △ : Biogas plants under construction or in project.

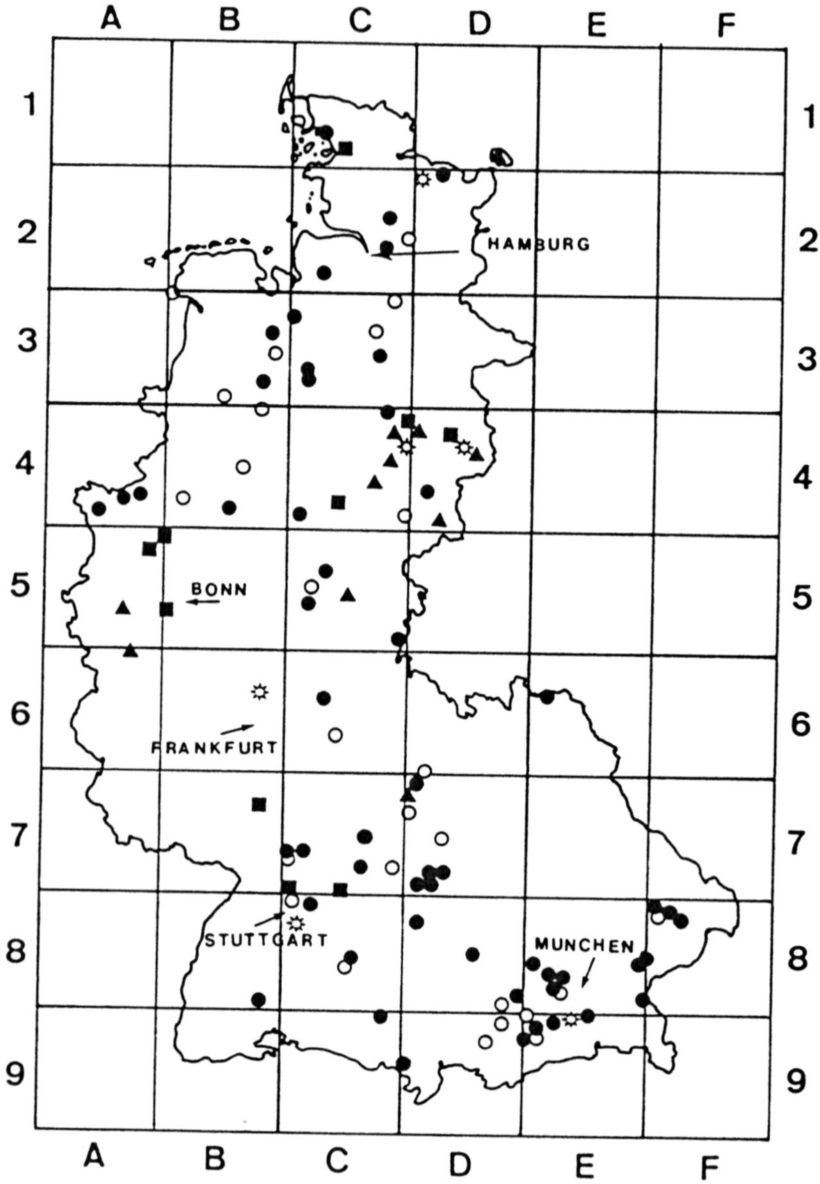

Fig. II.1.5. Location of biogas plants in the Federal Republic of Germany

● : Plants on stream on farm; ▲ : Plants on stream in industry; ■ : Landfills; O : Plants in project or under construction; ✡ : Research centres.

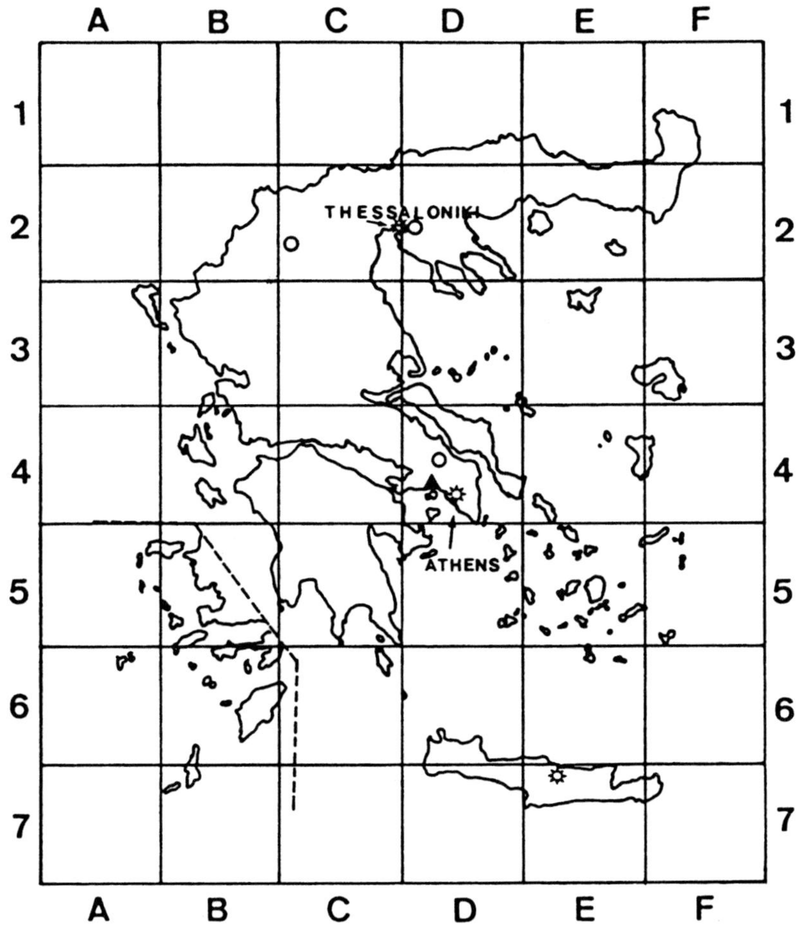

Fig. II.1.6. Location of biogas plants in Greece

▲ : Plant on stream in industry; O : Plants in project or under construction on farm; ☼ : Research centres.

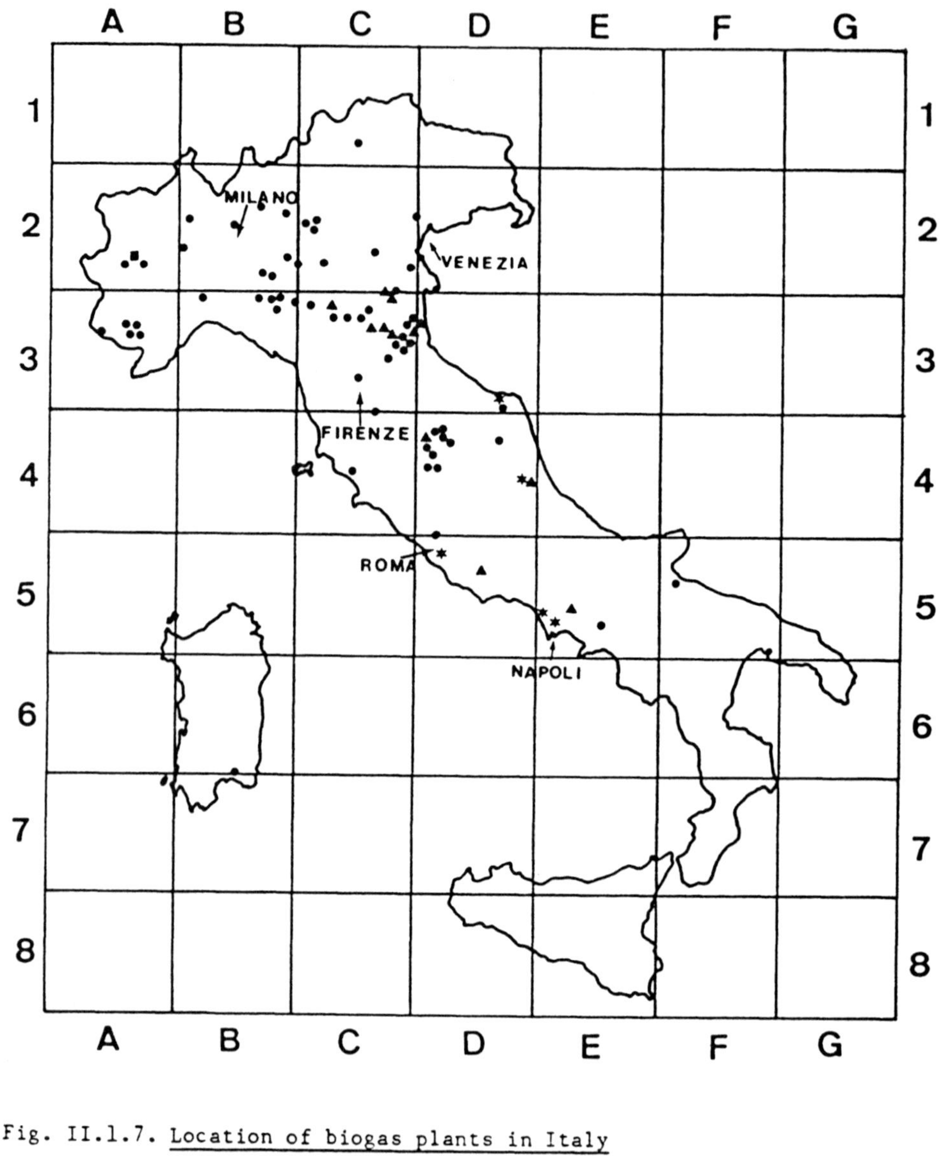

Fig. II.1.7. Location of biogas plants in Italy

• : Plants on stream on farm; ▲ : Plants on stream in industry; ■ : Landfill; * : Research centres

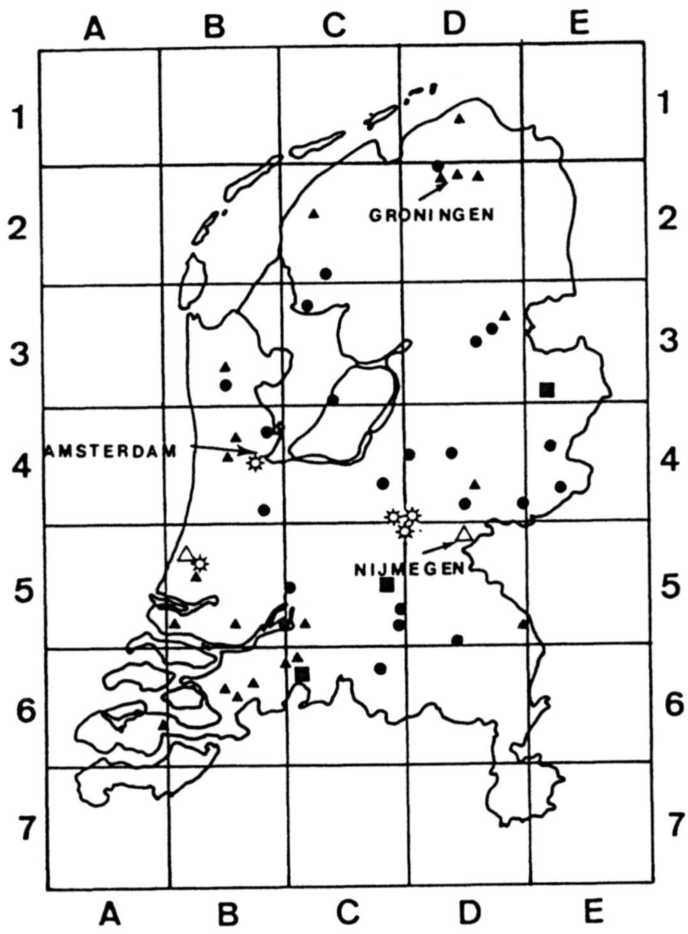

Fig. II.1.8. Location of biogas plants in The Netherlands

● : Plants on stream on farm; ▲ : Plants on stream in industry; ■ : Landfills; △ : Plants in project; ☼ : Research centres.

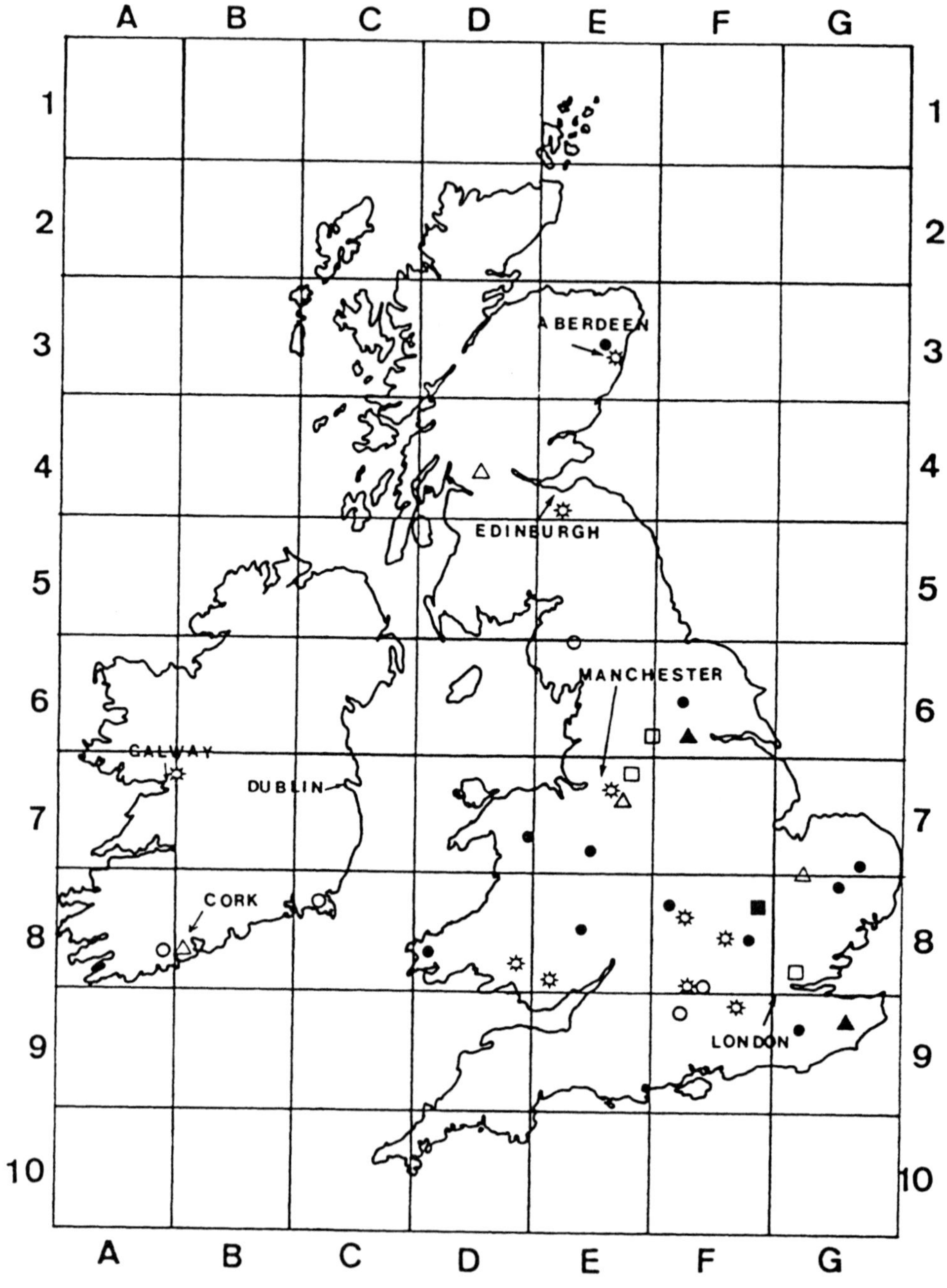

Fig. II.1.9. Location of biogas plants in the United Kingdom and Ireland

● : Plants on stream on farm; : Plants on stream in industry; ■ : Landfills; ○ , △ , □ : Plants in project or under construction; ☼ : Research centres.

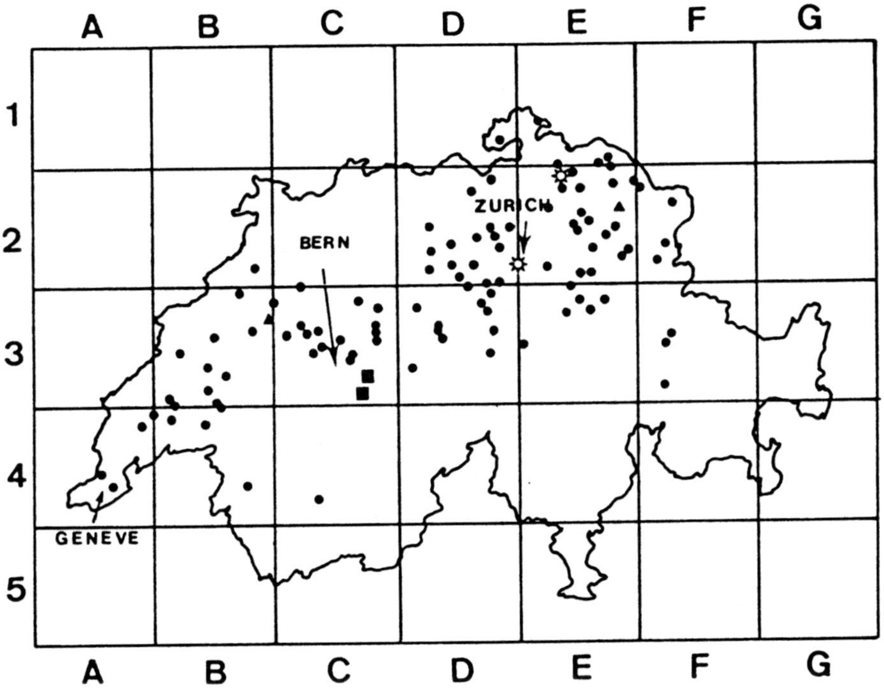

Fig. II.1.10. Location of biogas plants in Switzerland

● : Plants on stream on farm; ▲ : Plants on stream in industry; ■ : Landfills in operation; ☼ : Research centres.

digesters under construction started to increase significantly. Furthermore, as can be seen in Fig. II.1.11, the real proliferation of methane digesters on farms occurred after 1979-1980 and is still going on. The detailed evolution of the number of plants per year is available in Table 2.6 in Appendix II.B.

The effect of the energy crisis is evident, but it should also be recalled that the legislation on compulsory depollution of some agricultural wastes was enforced around 1979-1980. A typical example is given by Italy where the breeders of pigs have to pay for the pollution they produce.

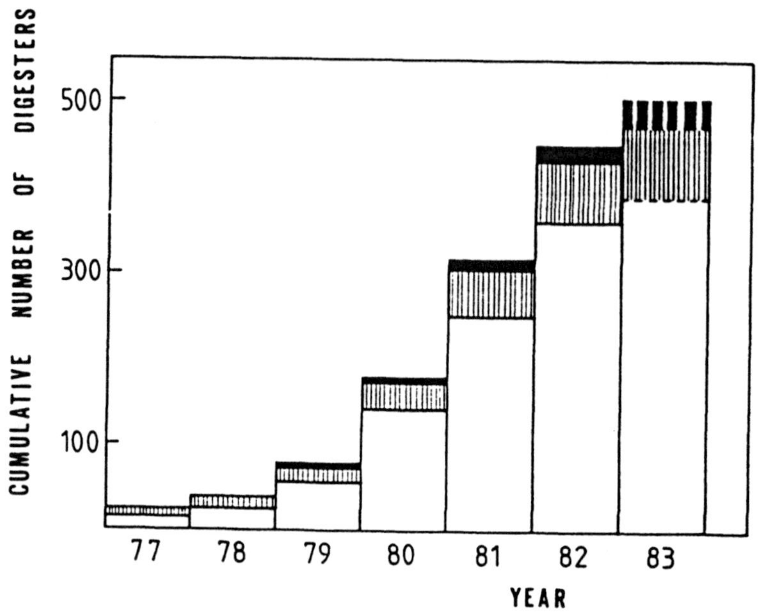

Fig. II.1.11. Cumulative number of biogas plants in Europe per year

▭ : Agricultural wastes; ▥ : Industrial wastes; ■ : Landfills

The biomethanation of industrial wastes was first applied in 1969 in Denmark and in 1970 in the United Kingdom. Fig. II.1.11 shows that the proliferation of industrial biogas plants started in 1980-1981. Since then, the number of methane digesters in industry has increased continuously.

As regards the abstraction of biogas from landfills, the first biogas plant was put on stream in 1979 in the United Kingdom. More landfills developed in 1981 in the Federal Republic of Germany and since 1982 in France and The Netherlands.

I.6. "Do-it-yourself" versus commercial methane digesters

Five different classes of constructors are interested in the construction of biogas plants. First, the owner of the biogas plant can also be the constructor (for example, a farmer builds his own full-scale agricultural methane digester). Secondly, a company may have the know-how and specialize in the construction of methane digesters. Quite a number of these specialized companies now exist in each country, and their addresses are listed in Appendix IV. Thirdly, local tradesmen can construct biogas plants, with or without a knowledge of biomethanation. The fourth and fifth classes are combinations of the three first classes, i.e. the owner with the help of a company or of local tradesmen.

Details of the distribution of agricultural biogas plants among the different classes of constructors is presented in Table 2.7 in Appendix II.B. It appears that, in most of the countries, full-scale agricultural digesters are generally constructed by a company. Only in three countries were less than half of the biogas plants constructed by a company. For Greece, this can be explained by the fact that there is as yet no specialized Greek constructor. One plant in Greece was constructed by an American company, the two other plants were constructed by local tradesmen. In France, it should be recalled that there are numerous batch digesters, especially the oldest ones, which were as a general rule constructed by the farmer himself. In Denmark, it seems that there is a preference for collaboration between the farmer and a company.

A relation can also be established between the mean sizes of the agricultural methane digesters and the different classes of constructors. Table 2.8 of Appendix II.B. details this relation. One might think that the methane digesters constructed by the owners themselves would be smaller than those constructed by companies. This is not always true, especially in France, The Netherlands and Switzerland. Italy is an exception : in this country, the average size of the methane digesters constructed by companies is 846 m^3, versus 189 m^3 for those constructed by the owners. Italy is also the only country where several agricultural methane digesters exceed 1 000 m^3 of working volume. The largest has a total volume of 7 000 plus 2 500 m^3. Pilot-scale methane digesters are mostly constructed by the owners, sometimes working together with local tradesmen.

Finally, the industrial biogas plants are in general designed by research departments and constructed by local tradesmen or specialized companies. Only four industrial companies in the Federal Republic of Germany have constructed their biogas plants themselves.

I.7. Motivations which led to the present status of biomethanation

The motivations of farmers or industrialists who have installed biogas plants were given in 103 answers. Only full-scale biogas plants were considered, since pilot biogas plants serve essentially for research and/or demonstration.

Table II.1.3 shows that the motivations for the installation of a biogas plant depend essentially on the type of waste to be treated. Biomethanation on farms is mainly employed for energy production or for energy combined with depollution. As might be expected, biomethanation of industrial wastes is always practised for depollution alone, or for depollution combined with energy production.

Table II.1.3. Motivations for installing biogas plants

Type of waste	Motivations (share of total in %)		
	Energy production	Depollution	Energy + depollution
Agro-waste	62	8	30
Industrial waste	0	63	37

Among the different types of agricultural wastes, depollution becomes a more important motivation in the case of pig manure (see Table II.1.4). For large pig breeding units, as in Italy, the reduction of the Chemical Oxygen Demand (COD) disposed of is an essential motivation, to which the benefits of a better odour control and water pollution control can sometimes be added.

Other side benefits of the biomethanation of agricultural wastes may be mentioned in some countries. Hygienization seems to be a side benefit in Denmark. The energy autonomy of the farmer seems to be important in Switzerland. The improvement of the quality of bedding manure after digestion can been mentioned in France.

Table II.1.4. Motivations for installing biogas plants on farms

Type of agro-waste	Motivation (share of total in %)		
	Energy production	Depollution	Energy + depollution
- Dairy cattle or bovine cattle manure (1)	94	0	6
- Pig manure alone or pig manure mixed with other manures (2)	42	11	47
- Poultry manure (3)	50	50	0
- Manure with bedding (4)	100	0	0

(1) 21 % of the answers concerned with agro-wastes;
(2) 61 % of the answers concerned with agro-wastes;
(3) 2 % of the answers concerned with agro-wastes;
(4) 16 % of the answers concerned with agro-wastes.

In the Federal Republic of Germany, a detailed inquiry was carried out by the national expert. Forty-four biogas plant owners were questioned as to their reasons for installing a biogas plant. The following motivations emerged from this inquiry : the production of energy, the reduction of odours (environmental protection), improvement of the fertilizer value of refuse, economic advantages when using the effluent as fertilizer, hygienization, the killing of weed seeds and the reduction of the amount of organic load.

The results of this inquiry are summarized in Fig. II.1.12. As was to be expected, the factors responsible for a positive decision varied from case to case.

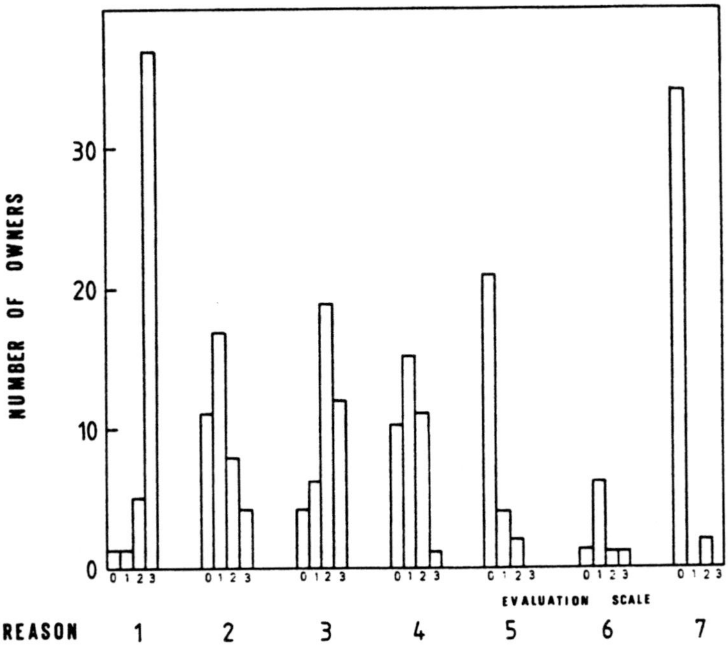

Fig. II.1.12. Evaluation by the owners of the reasons for installing a biogas plant

1. Production of energy;
2. Odour reduction;
3. Improvment of fertilizer value;
4. Economic advantages when applying the effluent as fertilizer;
5. Hygienization;
6. Killing of weed seeds;
7. Reduction of the amount of organic load.

The scale for the evaluation was : 0 point : no influence, 1 point : little influence, 2 points : fair influence, 3 points : great influence.

For most biogas plant owners, however, the most important reason was the possibility of covering a large part of their energy requirements with the gas produced. One point which is evaluated differently, depending upon the particular circumstances of each farm, is the reduction of odour expected from methane digestion due to the decomposition of the malodorous aromatic compounds in the liquid manure. The solution of the problem of odour connected with the application of the effluent as fertilizer can be of vital importance for many farms.

The second most important reason given was, in the view of the farmers, the expectation of being able to keep fertilizer costs on the farm down, thanks to the fermented liquid manure. Experts have found that decayed liquid manure, unlike normally stored untreated liquid manure, has no caustic effect on plants and thus can be spread on plants at the optimal period of nutrient requirement. Through this form of application, a considerable reduction in fertilizer losses can be expected, compared with the usual method of application in which the manure is spread on the bare ground.

In the past few years, farmers have observed with increasing concern a progressive and one-sided development in the variety of plants on intensively used fodder areas which are treated with mineral fertilizers or stored liquid manure. Grass varieties rich in leafy matter, growing near the ground, and nutritious meadow herbs were progressively suppressed, and plants with a lower nutritional value, such as high-stemmed and highly lignifying grass have tended to predominate. At the same time, the change from short-rooted to mainly long-rooted varieties of plants has caused large areas of the soil surface to be laid bare, so that, especially on the Alpine slopes, erosion damage has been increasingly observed, to the point of becoming a matter of vital importance for the farms located there.

Because decayed liquid manure is a non-caustic organic fertilizer, a large number of farmers, especially in the grassland regions, expect advantageous effects for soil organisms, soil structure and last but not least for the yield of forage and crop. If, under certain managerial peripheral conditions, these aspects are cumulated with the saving of fertilizer, then the increase in fertilizer value obtained by methane digestion as compared with all the other methods and their effects, can become the main reason for installing a biogas plant.

The economic advantage resulting from handling the decayed liquid manure and the timing of its application is the next reason, with a slightly lower degree of importance. Numerous farmers hope for alleviation of workload, thanks to the possibility of applying the non-caustic liquid fertilizer in a period of reduced activity following the cultivation period.

Furthermore, in farms where viscous wastes are produced, e.g. from bullocks, a reduction in work time is expected when homogenizing, transporting and spreading the digested liquid manure which becomes more watery during anaerobic decomposition. Farmers who partially use straw as bedding material in their stables are expecting also to achieve an easy-flowing end product through the simultaneous methane digestion of liquid manure and added solid manure, so that the additional handling and disposal of solid manure is no longer necessary. This would help to cut wages and machine costs.

In pasture farms and forage cropping farms, the hygienization effect is mentioned as an advantage of methane digestion. However, the scientific verification of this effect does not appear to have been sufficiently rigorous to satisfy some farmers and is accordingly not listed very high in their expectations. Only few farmers expect advantages from the feasible effect of killing weed seeds. No biogas plant-owner mentioned the reduction of the amount of organic matter in the liquid manure with a view to discharging the effluent to drainage ditches.

With a grain of salt, the foregoing analysis of the German farmers' motivations can be extended to all countries covered by the survey.

CHAPTER II. BIOMETHANATION OUTSIDE THE EUROPEAN COMMUNITY AND SWITZERLAND

II.1. Research and development activities in Scandinavia

The four Nordic European countries, Denmark, Finland, Norway and Sweden joined together in 1962 to set up a Nordic research cooperation called Nordforsk. This organization, which brings together national delegates of the four countries, is financially supported by the public authorities and has its main office in Helsinki. Nordforsk is not directly involved in the different national research programmes but aims at coordinating research within the four countries and the diffusion of the results obtained.

Within the Nordforsk association there is an Internordic Contact Group on biogas problems. This Internordic Contact Group is constituted by one delegate from each of the four member countries. The Group started its activities on 16 June 1981 and is committed for 4 years. At its first meeting in October 1981, the Group took the necessary measures to implement coordination of the different existing research projects on biomethanation in the four countries. In addition, they promoted the finding of more financial incentives for projects in fields related to biomethanation.

The first result was the sponsoring of an overview of the status of biogas plants on stream in each country. Secondly, four different complementary research programmes on biomethanation were launched. Each country is in charge of one programme. Status reports are sent to the three other member countries for advice and comments. The four research programmes deal respectively with :

1. Problems of heat exchangers and heat recovery (Finland),
2. Safety problems (Sweden),
3. Hygienization aspects (Norway),
4. Gas cleaning (Denmark).

The coordination effort of Nordforsk between the four Nordic countries has generated more collaboration. As an example, a study on the thermophilic digestion of excess sludges from wastewater treatment and of manures which had been carried out by the Danish company Cowiconsult and funded by the Danish Ministry of Education and the Laboratory of Technical Hygiene, was made available to the other member countries.

As well as coordinating research, seminars on biogas are being organized to promote the exchange of results and points of view. Three seminars bringing together about 40 participants have already taken place, one in 1977, one in 1980 and one in 1982.

II.2. Biomethanation in Spain

Spain has elaborated since 1978 a substantial programme on biomass conversion for energy purposes and environmental protection. The first studies established the potential waste biomass existing and available in Spain, based on a detailed survey in each province. The subsequent studies consisted in technical and economic evaluations of the various biomass conversion routes with a view to practical applications.

Biomethanation of waste biomasses is one of the conversion routes being studied at present. To date, the development of the technology of biomethanation has been limited to 18 methane digesters, of which 4 are industrial plants either in construction or start-up operation. Table II.2.1 details these different digesters and indicates the types of wastes treated, the volume of the digestion tank, the biomethanation system used, the location, the constructor and the promoter. Of these 18 methane digesters, 10 are in operation, 4 are under construction, 2 are in start-up operation and 2 are not running at all, one owing to technical problems and the other because the experimentation period has ended.

The wastes treated are in general wastewaters from agro-food industries such as slaughter-houses, distilleries and olive oil mills, or manure from bovine cattle or pigs.

The development of the technology of biomethanation is at present devoted mainly to the monitoring of demonstration methane digesters and to the economic evaluation of industrial digesters. The research itself is mostly oriented towards the design of second generation methane digesters.

The final objective is to gather by 1985 sufficient data on the evaluation of the various national and foreign technologies to gain sufficient knowledge of the structure necessary for their succesful implementation.

II.3. Status of biomethanation in the United States of America

The following paragraphs are the synthesis of a round trip in the United States of America which took place in June 1982, and should be taken as a European view of the U.S. status.

II.3.1. Summary

There is a general tendency in the United States to move away from biogas and anaerobic digestion as such or ethanol and fermentation as such or even energy as such and to consider as a priority goal chemicals in general from biomass through bioconversions in order to stabilize the world resource base in the long run. Within this general priority goal, however, depending upon local circumstances, individual sub-goals will make biomethanation important again, namely when considering integrated biogas-ethanol production systems.

With the exception of sewage sludge stabilization in municipal wastewater treatment plants, there is, at the present time, no really strong market in the United States for biogas plants. Motivations based on alternative energy supply, depollution of organic wastes, production of a compost-like

Table II.2.1. Biogas plants in Spain

Type of waste treated	Digester volume (m^3)	System National (N)/Foreign (F)	Location	Constructor	Financial promoter
		I. Demonstration plants			
Paper industry wastewater	5	Anaerobic contact (F)	SAICA-Zaragoza	AC Biotechnics	SAICA
Manure	10	Completely mixed (F)	CCA-Soria	Hamworthy	ETSIA
Cattle manure	15	Completely mixed (F)	INIA-Badajoz	Simplex	DGPA-INIA
Pig manure	15	Anaerobic contact (N)	IGD-Sevilla	ENADIMSA	ENADIMSA-DGPA-INIA
Pig manure	18	Completely mixed (N)	ETSIA-Madrid	Aplesa	ETSIA
Pig manure	20	Anaerobic filter (F/N)	Farm-Soria	Sufisa	DGPA-ENADIMSA
Landfill leachate	40	Anaerobic contact (F)	Vertedero-Barcelona	Cida-Hidroquim	Province
Plant material	50	Completely mixed (F)	CRIDA 06-Madrid	Simplex	DGPA-INIA
Bovine cattle manure	50	Batch (N)	Farm-Gerona	ENADIMSA	Province-ENADIMSA
Olive oil mill waste-water	70	Anaerobic contact (F)	IGD-Sevilla	Biomecsa	IGD-CSIC
Bovine cattle manure	100	Completely mixed (N)	CRIDA 03-Zaragoza	De Gaspari	DGPA-INIA ENADIMSA
Pig manure	100	Anaerobic filter (F/N)	Farm-Toledo	Sufisa	DGPA-INIA-ENADIMSA
Pig manure	140	Completely mixed (N)	Farm-Barcelona	Catalana Gas	CG-IRYDA-CEE
Distillery wastewater	250	Anaerobic contact (F/N)	CIASA-Sevilla	Ateinsa	ENADIMSA

Type of waste treated	Digester volume (m^3)	System National (N)/Foreign (F)	Location	Constructor	Financial promoter
		II. Industrial plants			
Pig manure	500	Anaerobic contact (N)	Carcesa-Mérida	ENADIMSA	DGPA-INIA-ENADIMSA
Pig manure	500	Completely mixed (N)	Carcesa-Mérida	ENADIMSA	DGPA-INIA-ENADIMSA
Bovine cattle manure	1 150	Completely mixed (F/N)	Tauste, SA-Zaragoza	Casis, SA	DGPA-INIA-ENADIMSA
Slaughter-house wastewater	1 400	Anaerobic contact (F/N)	Frigsa-Lugo	INITEC	CEE-INITEC-Frigsa

Abbreviations employed in Table II.2.1.

CARCESA : Empresa Carnica
CCA : Centro de Capacitacion Agraria
CEE : Centro de Estudios de la Energia
CG : Catalana de Gas
CIASA : Compania de Industria Agricola Sociedad Anonima
CRIDA : Centro Regional de Investigaciones y Desarrollo Agricola
CSIC : Consejo Superior de Investigaciones Cientificas
DGPA : Direccion General de Produccion Agraria
ENADIMSA : Empresa Nacional Adaro de Investigaciones Mineras Sociedad Anonima
ETSIA : Escuela Tecnica Superior de Ingenieros Agronomos
FRIGSA : Empresa de Produccion y Derivados Carnicos
IGD : Instituto de la Grasa y sus Derivados
INIA : Instituto Nacional de Investigaciones Agrarias
INITEC : Empresa Nacional de Ingenieria y Tecnologia
IRIDA : Instituto Regional de Desarrollo Agricola
SAICA : Empresa Papelera
TAUSTE SA : Empresa de Explotacion Ganadera

fertilizing soil conditioner, are weak. Interest rates of money loans are high. Specific incentives from public authorities are inexistent.

There is a large potential market in the United States for bioconversion systems in general and for methane digestion systems in particular. A general aim as defined above does exist. Motivations, also defined above, if weak, do exist. Any combination of two of these motivations already makes existing anaerobic systems economical, competitive if not attractive. No one can however foresee when this potential market will turn into a real market.

As a result of these three conflicting points, some research, development and demonstration in the field of biomethanation is going on in the U.S. Transfer of this knowledge towards industry and popularization of this knowledge towards potential clients is weak. Few private companies are involved with biomethanation. The United States no longer has a visible lead in the field of biomethanation; this visible lead now appears to be in the hands of Europeans. However, multinational companies are presumably preparing themselves in view of the general aim defined above, although little evidence can be brought forward to support this assumption.

Existing research and development programmes point to the believed priority in the United States of the necessity to improve the knowledge of the basic phenomena involved in bioconversion processes so as to develop the principles of second generation technologies rather than to demonstrate existing mature technologies, although the latter is being carried out albeit in a rather sporadic way.

II.3.2. United States policy in the matters of biomass, bioconversions and biomethanation

The general policy of the present Administration is that public money should finance long range, high-risk research with high potential payback in the future. What is allowed is to test the concept and this axiom limits the largest experimental scale allowed. The latter does not usually permit to state the economic feasibility of the process.

The present U.S. Administration is strongly pro-nuclear. Reasons for this are obscure but may involve national defence. In the United States, nuclear energy represents 2.7 Quadrat British Thermal Unit (Q BTU)[(1)], a figure to compare to hydroelectric power, calculated as 2.9 Q BTU and the present uses of energy from biomass, estimated at 2.5 Q BTU, each representing roughly 3 % of the total U.S. energy supply. As a result, most of the research money goes to nuclear R, D & D, although there are no new orders for nuclear power plants, presently believed to be too expensive.

The present U.S. Administration is down-plating renewable energies. Research in Energy from Biomass, managed by the U.S. Department of Energy (DOE) represented 46 M$ in 1981, 20 M$ in 1982, and is expected to drop again in 1983. That is, provided the United States Congress follows the present Administration, which is unlikely to happen.

(1) 1 BTU = 1.055×10^3 J

One of the reasons why biomass is ignored by the present U.S. Administration is that biomass is captive, i.e. it is not a marketed product, namely the producer is the user.

The present U.S. Administration wishes to withdraw the control of oil prices. This control, as well as the control of gas prices is in the hands of the Department of Energy. The weakening of the Department of Energy, consequent to this move, is destabilizing the research policies within the Department of Energy.

II.3.3. Targets of biomethanation in the United States

In the short term, wastewater treatment is a possible target for biomethanation in the United States. Whey is a notable example : in the United States, whey is a waste because most if not all the cheese is made in only three States, a fact which severely limits local reuses of whey. Whey can become a feedstock for both ethanol and biogas production. However, large companies are interested only in a full-scale, well demonstrated biogas plant, and do not want to be involved in research and/or development.

There are in the United States large amounts of ligno-cellulosic and herbaceous wastes. Ligno-cellulosic materials, such as stalks or straw, may not be suitable for thermoconversion processes, because the feedstock is scattered too widely. The latter processes, to be economical, require large quantities of feedstock and the gathering of the feedstock is too expensive. In contrast, bioconversions are globally as economical in small- as in large-scale units. Ligno-cellulosic materials may also not be dry enough for economical thermoconversion. Above 50 % moisture content, it is believed that bioconversion processes become more interesting than thermoconversion processes. Besides, materials containing large amounts of pectines are less suitable for ethanol fermentation than for methane digestion.

Farms and their agricultural wastes are the classical target, in the United States as elsewhere, provided energy production can be associated with depollution. Certain industries become targets for biomethanation in as much that the field of biomethanation is widened to include a larger number of low cost resource compounds. The idea is to stabilize the price structure of basic compounds in the long term. These industries hope to appropriate raw materials and the production market for these raw materials.

Targets of biomethanation in the United States may be more easily understood when considered together with incentives and motivations, to be discussed below.

II.3.4. Existing biogas plants in the United States

The preceding paragraphs do not suggest a large number of full-scale biogas plants in the United States. Indeed, not more than 50 biogas plants in operation can be located at present in the United States and are scattered over the whole country.

Biogas projects, however, had started to be funded by Government contracts in the United States around 1972. At that time and later, some biogas plants were put up with the aid of the Energy Research and Development Agency (ERDA). "Anaerobic Digesters in United States" compiled by Anaerobic Energy Systems, Inc. (Bartow, Florida) and issued by the U.S. Department of Energy (Washington) lists 23 biogas plants and appears as a good status in 1980.

Other biogas plants appeared in a disorganized way, e.g. as demonstration units for a commercial firm with or without local granting. The first commercial demonstration biogas plant was built in 1978 in Brawley, Southern California. It was running on manure and operated by the Pacific Gas and Electric. For motivations similar to those elicited by the survey in Switzerland and detailed in Part II, Chapter I, a number of isolated ventures took place in the Midwestern United States : Minnesota, Wisconsin and Iowa. In Florida as well as in California, some biogas plants are running on water hyacinths grown for wastewater treatment.

More recent commercial biogas plants, mostly if not all on farms, mostly if not all on animal manure, have been identified by Hashimoto (1983).

In the general field of biogas, commercial firms known to be dealing with biomass, energy from biomass or bioconversions are often small companies, presently struggling to survive, at least in the particular field of biomethanation. It is estimated that between 10 and 15 companies are presently marketing traditional biogas plants, that is including a methane digester of the completely mixed or plug-flow type.

II.3.5. Research in biomethanation in the United States

Whereas basic research in biomethanation is being funded (to a minor extent) by the National Science Foundation, it is the duty of the U.S. Department of Energy (DOE) to finance applied research in biomethanation. This is being done mainly by its Biomass Energy Technology Division, to a certain extent by its Biological Energy Service and for a particular aspect by its Municipal Solid Waste Programme.

The Solar Energy Research Institute (SERI) has the lead in applied research in biomethanation in the United States. It is being funded by the U.S. Department of Energy and coordinates for the latter all research projects funded within its Biomass Programme.

SERI's scientific policy has moved from the demonstration of existing mature technologies : methane digestion of residues, mainly of agricultural origin, mainly animal manure, as originated in 1972 by the Energy Research and Development Agency towards first, improved knowledge of basic processes, secondly, expanded product opportunities. From 1 January 1982, existing demonstration programmes within the U.S. Department of Energy granting scheme will only be awarded grants for their completion. No new demonstration programmes will receive grants.

It is at present the duty of the U.S. Department of Agriculture (USDA) within its Agricultural Research Service to demonstrate the commercial feasibility of large-scale methane digesters loaded with animal manure or crop residues. The programme has however a much broader scope and involves biomass as a source of energy and chemical feedstocks.

East Grand Forks (U.S.A.). 28 000 m^3 anaerobic contact methane digester on sugary wastewater.

Bartow (Florida, U.S.A.) 1979. Twin 6 000 m^3 completely mixed continuous methane digesters on cattle manure.

Research programmes are also being funded by local authorities such as the California Energy Commission, the Coral Spring Improvement District of Florida or the Minnesota State Department for Energy Development. Other research programmes are being funded by private institutions such as the Bio-Energy Council (Washington, D.C.) or the Gas Research Institute of Chicago through its Institute of Gas Technology (IGT). The best known project is the biogas plant on water hyacinths grown for wastewater treatment at Disney World (Orlando, Florida).

II.3.6. Transfer from ideas to technology downstream from University to Industry in the United States

The present U.S. Administration expects the transfer of university ideas to industrial technology to take place by itself, without incentives, and in a short term (5 years). If a new energy product is proposed as a result of R & D subsidized by public authority funds, industry will rapidly make a marketable product from this. This axiom is proven by the following examples. R & D on Euphorbia has led the chemical industry to produce resins to be used in place of latex. R & D on micro-algae has led the chemical industry to market special chemicals. R & D on alcohol fermentation has led General Electric to obtain lignin which, by hydrogenation, yields marketable phenols. Furthermore, the delignified straw constitutes a fine pure feedstock for high-grade paper or cellulosic membranes, in as much that these byproducts have taken over and that ethanol is no longer produced.

The ideas of the present U.S. Administration are not generally followed; among the reasons for this are the two developed below.

Whatever comes out of research subsidized by U.S. Government funds is community property. This output of research can be protected by U.S. Government patents. But these patents can be used by any U.S. industry. This means that a U.S. industry which invests money and time to develop an economical process or a marketable product from a U.S. Government patent does not receive any protection for doing so. It does not have to pay royalties either. Such a system is believed to be improperly constructed because it leads to a situation where industry no longer wants to be involved in subsidizing in part university R & D.

There is a tendency, the strength of which cannot yet be evaluated, to modify this situation by scrutinizing what is being done in Japan, France, Germany and Israel with the aim of proposing some sort of exclusivity, to be defined, to protect an industry interested in developing a U.S. Government patent, in exchange for some form of royalty, also to be defined.

Under former President Carter's Administration, a number of guaranteed loans had been given to various commercial firms to test and demonstrate biotechnologies, mainly in the field of ethanol fermentation. Under the present U.S. Administration, these loans may no longer be guaranteed. Hence private companies now have "cold feet" and are no longer willing to invest in the development of new technologies arising from universities or the public research world.

Ignoring, or better, not taking part in the R, D & D of renewable energies, some major industries have moved into the biomass field in simple ways but very strongly. An outstanding example is the wood, pulp and paper industry, a major energy consumer. Since the oil crisis in 1973, woody wastes have been used to produce energy, mainly by combustion in appropriate furnaces. The investment for these furnaces amounted in one case to 80 M$, but the investment was paid back in 2-3 years. Presently the wood, paper and pulp industries are 60 % energy self-sufficient and expect to become 100 % energy self-sufficient. There are two motivations for this move. First, for these industries, energy from biomass is cheaper than energy from other sources. Secondly, energy independence appears to be an economically attractive situation. Other types of industries, where feasible, are attracted by this situation.

Similarly independently of public research, the private sector apparently does make some efforts to improve biomethanation. Little information is available. Patent protection is sought. But funding sources such as the Bio-Energy Council, without being allowed to give details, claim that ideas leading to significant advances in the general field of bioconversions are being submitted for funding.

To promote contacts between University, Administration and Industry, a Biomass Energy Research Association (BERA) was started early in 1982. It is a non-profit educational association whose aim is to lobby in the biomass energy field. It is habilitated to testify before Congress.

II.3.7. Popularization of biomethanation in the United States

Little is organized by the present U.S. Administration to popularize the idea of biomethanation in the United States. This should be the duty of the Department for Commerce through part takings in trade shows. More seems to be achieved in fields such as photovoltaics and windmills.

Notwithstanding the preceding paragraph, isolated trials for the popularization of biomethanation have been identified. In Arkansas, the Arkansas Department of Energy, located in Little Rock, is diffusing two booklets written by Indigent Energy Associates, Inc., and Perennial Energy Inc. The first is a "Guidebook for Small Scale Methane Facilities". The second concerns "Technology Assessment of Anaerobic Digesters for Arkansas Farms".

Clear-cut demonstrations in the field of biomethanation are universally desired by research, production and administrative technologists. However, the prospect for a nation-wide organization of clear-cut demonstrations of biomethanation units appears to be poor in the United States, as presently no U.S. entity appears involved or even interested in doing this. It should be the responsibility of industry, but few industrial organizations if any prospect for this new potential market. The structures which should popularize and diffuse (industrial marketing and applications) have a conservative stand.

II.3.8. Economics of bioconversions in general, of biomethanation in particular in the United States

With regard to biomethanation, there are claims in the United States (*e.g.*

by Clausen and coworkers in Gaddy's group at the University of Arkansas) that biogas plants are economical per se. This is still questionable.

However, with added values (fringe benefits), the methane digestion system looks definitively and definitely economically attractive. Among added values, one finds the use of the effluent from methane digestion for animal feed. This raises a problem particular to the United States. Whereas the use of the effluent from methane digestion for animal feed usually does not raise problems within any of the U.S. individual States, where local regulations can more easily be adapted, this is not the case when this added value involves interstate business, which falls under Federal regulations and where clearance for this added value is not yet gained. The situation might be cleared within the policy of the present Administration to minimize Federal regulations and to strengthen State regulations.

One of the economic drawbacks of biogas for pipeline distribution is the costly necessity to upgrade biogas by removing sulphydric acid, H_2S, and carbon dioxide, CO_2. A cost study, performed on biogas abstracted from landfills in California has shown that it is more expensive to clean biogas than natural gas. Therefore, biogas economics limit its use to local ones using modified burners, a limit which does not hamper the economic attractability of the biomethanation of agricultural and municipal wastes.

Care should be taken to avoid premature definitive conclusions. The example of ethanol is striking in this respect. In the United States, ethanol by fermentation has just become cheaper than ethanol from ethylene. Already sales contracts for 30 Mgallons have been passed with industry, notwithstanding the 70 Mgallons of ethanol used as gasohol, where tax incentives are still necessary at the level of 4 U.S. cents per gallon.

II.3.9. Motivations and incentives in the United States

The farm situation, particular to the United States, favours biomethanation and, hence, elicits a large but hitherto potential market. First, U.S. farms are large-scale, industrial-type agricultural exploitations, a situation generally less frequent in Europe. Secondly, U.S. farmers are innovative, hence ready to test attractive technologies. Thirdly, in some cases, excess electricity, produced from biogas, is being sold back to electricity companies, although no further details or figures were available. Finally, energy farms do start to appear in the United States. In these farms, grain crops are used for alcohol production. The fermented mash after ethanol distillation is used for pig feeding. The pig manure is used to produce biogas which is used as energy supply for ethanol distillation. Ethanol is used as liquid fuel for tractors and cars.

At present, the incentive of depollution in farms concerns mainly odour problems and arises as the cities increase in size and enclose farms.

There are few financial incentives. First, there are no particular subsidies in the United States for building biogas plants. However, there are general financial incentives applicable to the extension of small business firms. These general financial incentives are often applicable to biomethanation. As an example, in the United States there exist governmental unds for small businesses, universities, and non-profit organizations to develop patents they have applied or paid for.

The reasons why industry has not yet taken off in the field of biomethanation are, first, that there are no real reasons per se to invest in biomethanation, secondly, that wastes can still be thrown away, and thirdly, that presently the existing resources of energy remain sufficient. There is a need for liquid fuel but not for pipeline distributed biogas. There is enough natural gas in the United States of Mexico, the United States of America and Canada.

II.3.10. Legislation and safety regulations in the United States

There is little information on particular regulations in the United States. As far as biomethanation itself is concerned, there is no particular U.S. legislation in the field.

As far as outputs are concerned, in the United States electricity can be exchanged between companies as part of trade services. It can now also be sold to electricity utilities at cost price which is between 25 and 30 % of sales price; this has occurred *e.g.* in Bartow, Florida. Is it economic ? The answer is no, but what can you do if you have excess power !

Details of U.S. research programmes, a list of addresses and a bibliography will be published in the *International Journal of Solar Energy*.

CHAPTER III. INCENTIVES TO PROMOTE BIOMETHANATION WITHIN THE EUROPEAN COMMUNITY AND SWITZERLAND

III.1. Incentives by legislation on depollution

a. Belgium

As far as agricultural wastes are concerned, there exists a basic legislation on the limits of pollution but until now, the legislation has not been strictly enforced. Agricultural wastes are disposed of on land, of which the available area is normally sufficient to avoid water pollution with the exception of the north-western part of Belgium where there are intensive piggeries. As regards industrial wastewaters, the Royal Decree of 22 April 1977 determines the sectorial conditions for discharge, but the application of this decree has been little enforced.

b. Denmark

The Environmental Protection Act no. 372 of 13 June 1973 has amongst its objectives the prevention of pollution of air, water and soil. It also provides the basis for the planning and implementation of pollution control. Both agricultural and industrial wastes come under this law.

c. France

The limits of pollution in France are fixed by local public authorities called "Agences de Bassin" which are responsible for reducing the pollution caused by industries or cities. They can tax polluters and fund investments aimed at limiting pollution.

d. Federal Republic of Germany

The regulations of the Federal Law on Pollution Protection aim at protection against unreasonable emission with a view to reducing danger, as well as precautions against harmful environmental effects and organized waste disposal. Precautions must take into account the current state of technology in pollution control. "The state of technology" is therefore understood as being a dynamic expression. There is a specific need for authorization, according to this law (in addition to building permission), if treatment of residues in biogas plants concerns large piggery or chicken breeding units. In these cases, the technical instructions for keeping the air clean are also applicable. These technical instructions as well as the DIN (Deutsche Industrie Normen) guidelines nos. 3471 and 3472 contain technological measures for storage tanks and storage times, manure transport, etc., as well as distance recommendations. The statutory duty of registration with sewage authorities is also applicable to farms with excess manure.

The Federal Water Resources Law deals with all forms of waste disposal including, according to the legal definition, the dehydration of sewage sludge. Sewage is mentioned explicitly in administrative regulations, as a result of this Federal Law, with minimum requirements for the innocuity of domestic and communal foul water. The Sewage Use Law can have immediate effect on technical plans.

The Federal Waste Disposal Law only affects biogas plants if communal or industrial sewage sludge is to be disposed of, or purified liquid manure, sewage sludge or similar substances are produced as fertilizer and, in the case of animal waste, the usual amount of agricultural fertilizer is exceeded. The basic disposal requirements of the Federal Law are partially based on criteria similar to the main requirements of Building Law. A recent amending law deals with its "use", alongside the "production" of agricultural matter.

The use of sewage sludge in agriculture in a way which is harmless from an epidemic point of view has up to now been dealt with individually in each State. Since April 1983, there is a new legal ordinance for the whole Federal Republic which provides for an environmentally and sanitarily acceptable use of sewage sludge. Furthermore, the supply of drinking water must be considered.

The Federal States have issued State Water and Waste Laws in addition to the Federal Environmental Laws. Within the States, there are also Security and Police Regulations about the transportation and production of liquid manure within their boundaries. Application regulations with a view to avoiding over-fertilization are locally in preparation. Corresponding authorizations for intervention by the Government are being studied.

e. Greece

The Greek legislation concerning limits of pollution from urban and industrial liquid wastes is strict for surface water pollution requirements which include among others, a minimum dissolved oxygen content requirement of 3-5 mg/l and a pH 6 - 9.5 depending on the water use. For land application of wastewaters, legislation allows higher limits but only for good quality soils that are well drained and managed to accept wastewaters without over-saturation problems. New legislation concerning livestock and poultry wastes is under preparation.

f. Ireland

Effluent discharge in Ireland is governed by the Local Government (Water Pollution) Act of 1977. The controls specified by the act prohibit the discharge of trade effluents and certain sewage effluents to waters except in accordance with a licence. The licence is obtainable from the Local Authorities (county, borough and city authorities) and is granted subject to a variety of conditions relating to the effect of the discharge on the receiving water, the method of treatment, the taking of samples, etc. The licensing requirements do not apply to discharges from vessels or marine structures to tidal waters or to the sea and discharges from public sewers are also exempt. The Minister for Environment has power, however, to prescribe standards for such discharges.

A licence is subject to review at intervals of not less than three years. It may, however, be reviewed at any time should the authority believe that the discharge is a significant threat to public health or where an unforeseen material change has taken place in the condition of the receiving waters. An appeal is allowed under the act by any person in relation to the granting or refusing of a licence by a local authority or to the conditions attached to such a licence. All such appeals are directed to a Bord Pleanala (National Planning Board). The board may refuse the appeal or give appropriate directions to the local authority concerned.

Prosecutions in respect of offences against the act may be brought by a local authority, a board of conservators, the Minister for Fisheries or by any other person. The Act, in addition to providing for prosecution, also empowers the local authority, board of conservators or the Minister for Fisheries to apply to the District Court for an order directing the person concerned to mitigate or remedy the effects of a contravention of the licensing requirement.

The Act empowers the Minister for the Environment to make regulations prescribing standards for waters and for effluents or for methods of treatment of effluents. In general, the standards are effectively Royal Commission Standards.

g. Italy

According to the "Merli" law (10 May 1976) no. 319, breeding units (piggeries or cow houses) are classified as civil or industrial types depending on whether they have less or more than 4 000 kg of weight of live cattle per hectare of land. The industrial breeding units were compelled to comply with the provisions of the law before September 1981, which means either use of enough land or treatment of the manure.

According to law no. 690 issued on 10 October 1976 the main regulations for industrial wastes and for industrial breeding units are the following :

- the depollution programme had to be submitted to the local public authority (Province and/or Commune) before 29 February 1980;
- the treatment plant had to be completed not later than 1 September 1981 (the date limit was extended afterwards);
- starting from 30 December 1979 up to the completion of the plant, the breeder or industrialist must pay for the pollution produced. This sum is related to the cost (C) of the projected treatment plant and is calculated by $[(7/12) \times C \times N]/100$ where N represents the number of months elapsed from 30 December 1979.

Civil breeding units which do not spread all the manure on land are allowed to discharge wastewaters into the public sewer system, where it is available, at standards determined by Local Public Authorities. The main regulations of the law for civil wastes are the following :

- Law no. 690, item 1 (8 October 1976) provides a further specification of agricultural exploitations as "Civil Installations";

- Law no. 650, item 17 (24 December 1979) specifies the Public Administrative Body which is consulted about the civil or industrial nature of agricultural exploitations;
- Law no. 901 (30 December 1980) fixes the costs of the collection of the wastewaters at 20 LIT (0.015 ECU)/m^3 and of the treatment of the wastewaters in a public sewage plant at 20 LIT (0.015 ECU)/m^3. The volume of the wastewaters is assumed equal to 80 % of the fresh water consumed on the farm.

h. The Netherlands

The principal objective of manure digestion in The Netherlands is the production of energy to cover the main part of farm energy needs. Another objective, which will be increasingly important in the near future, concerns the abatement of malodour nuisance at intensive animal production units.

The disposal of domestic and industrial wastewater is regulated by law, which imposes treatment of the water. Direct anaerobic treatment of wastewater offers significant benefits over the commonly applied aerobic process with respect to energy demands, the presence of nutrients, the production of excess sludge and the building area required. Therefore, anaerobic treatment often competes successfully with aerobic treatment processes. This is already the case for most medium-strength and concentrated wastewaters containing mainly organic material.

i. United Kingdom

Pollution control is seen as one of the main incentives for installing a methane digester apart from the energy value of the biogas. Water pollution control has however been less effective as an incentive on farms than it has been for methane digestion of sewage sludges and industrial wastes. The main reason for this is that farmers have not been pressurized to any great extent to implement controls.

In practice, most intensive livestock effluents will be disposed of to land and, if carried out in the approved way, should not present any problem. Water authorities do not have to be informed of land discharges, but, if there is an accidental run-off into a watercourse, they are entitled to prosecute. On the other hand, air pollution in terms of odours is a greater incentive to install digesters. Odour coming from intensive animal units and collection of dung can be a public nuisance. If they receive complaints, local authorities have the power to shut down a farm until control measures have been taken.

j. Switzerland

Within the framework of the Federal Law on Water Protection of 8 October 1971, the Order of 8 December 1975 concerning the discharge of wastewaters forbids the dumping of agricultural effluents into watercourses. They must be used on land in such a way that they do not pollute underground and superficial waters. In view of this requirement, the spreading of agricultural effluents is not allowed during wintertime and Swiss farmers are obliged to store the effluents during this period.

III.2. Incentives by legislation on the sale of gas and electricity

In most countries, the sale of gas or electricity is controlled in a monopolistic manner. Hence, when the central companies accept to buy the gas or electricity produced by a biogas plant, they do not pay a high price for it. The situation concerning the sale of gas and electricity has been analysed in more detail for three countries : Denmark, Italy and the United Kingdom.

a. Denmark

There is no legal or practical obstruction to the sale of surplus electricity to the public grid. The price for the electricity sold is usually approximately 75 % of the price paid for electricity. The price can, however, vary from one part of the country to another as the price is fixed by the local electricity company.

Special rules stipulate that the electricity companies must accept the surplus electricity produced and must pay a price which is at least equal to their own savings in running costs.

Biogas plants producing less than 75 kW electricity are not subject to energy taxes. The current energy tax is 0.125 DKR (0.016 ECU) per kWh and the Value Added Tax (VAT) is 22 %. Energy used on the farm and the plant is free of energy tax and VAT.

b. Italy

Electrical energy in Italy is almost totally supplied by ENEL (Public Body for Electrical Energy). In November 1981 this body issued standards for the connexion of TOTEM (Total Energy Modules) to the ENEL system. These standards take the following aspects into consideration : first, the installation of TOTEM in parallel with the ENEL electric system, secondly, guidelines for electrical safety precautions, as requested by ENEL; thirdly, regulations for the parallel operation of a TOTEM electric generator with the ENEL electric system; and finally, the application of CIP (Ministerial Committee for Price-list) provision no. 45 (22 October 1979). As from the end of 1981, ENEL will pay 40.5 LIT (0.031 ECU) for each kWh of electric energy produced by TOTEM.

c. United Kingdom

Until recently, the sale of gas in the United Kingdom was controlled by British Gas in a monopoly situation. However, legislation has been passed which allows other large business concerns which produce or trade in natural gas to also sell it for the purpose of energy generation. The new act contains sections which permit the sale of biogas by private producers to large consumers. New legislation is being enacted to allow privately generated electricity to be sold to the national grid, provided that certain safety and technical aspects are complied with. Prices to be paid by the local electricity councils will be based on the Central Electricity Generating Board (CEGB) basic tariff but taking into account service and extra investment costs required to accept such surplus electricity. In practice this may not prove to be economic but it will at least make it legally possible to sell electricity.

III.3. Financial incentives

a. Belgium

Subsidies for farmers wishing to install a biogas plant are available through the Ministry of Agriculture and Economic Affairs (Belgian Official Journal of 4 September 1981). Farmers must first submit a request to the Ministry of Agriculture. If approved, 35 % of the investment cost of the biogas plant is reimbursed. The cost of a one-year monitoring of the biogas plant is also covered.

Grants for applied research on biomethanation have been provided for by the IRSIA (Institute for Scientific Research in Agriculture and in Industry) through the COVAL programme (Research Committee for the Valorization of Residues in Agriculture) and through the COMET programme (Research Committee on Biomethanation, which includes seven industrial companies). Other grants for research have been provided for by local or regional authorities (Province de Brabant, Région Wallonne, Vlaams Gewest) and by the Commission of the European Communities.

b. Denmark

The availability of public support for the installation or the development of biogas plants has fluctuated over the past few years. In general, there is a political will to support both the research in and development of alternative energy resources, and also a will to support the installation of biogas plants. However, with regard to the installation of biogas plants, it is a general policy only to use the support as an incentive at the beginning of a development ("to prime the pump") for investments which are economically attractive from a social point of view, but which are not so from a private point of view.

The total amounts available in the future for support of biogas plants and the specific rules to be applied are therefore dependent partly on the development of the biogas economy and partly on the general political development and the resulting economic "package deals". At present, subsidies may be obtained from the following sources. Support for research can be obtain firstly from the Ministry of Energy and cover partially or fully expenses based on individual applications. Funds are allocated on a 2-3 year basis. Secondly, they can be obtained as general research funds for development of technology and agriculture (in competition with other research activities).

Support for the installation of biogas plants can be obtained, first, from the Ministry of Housing. A subsidy of 25 % can be obtained if a biogas plant of an approved type is erected on a commercial basis (standard application procedure to be followed). It can also be obtained from the Ministry of Environmental Affairs. A subsidy of 25 % can be obtained for process alterations which minimize external pollution, provided that the arrangements can be categorized as an economic burden. Economic benefits from the installation are deducted before evaluation (individual application). A subsidy of 25 % can be obtained for experiments concerning new technologies which reduce external pollution (individual application). Generally, deductions are made when funds are obtained from more than one public source.

c. France

There are seven sources of subsidies for biomethanation projects in France. First, the Agence Française pour la Maîtrise de l'Energie (AFME) has a budget for 1982 of 11 million French francs (1.77 MECU) which includes 1.4 MF (225 kECU) for fundamental research, 3.7 MF (600 kECU) for calls for the "Digesters on Farms" proposal, 2.5 MF (400 kECU) for the follow-up of demonstrations, 1.3 MF (209 kECU) for socio-economic studies, and 2.1 MF (338 kECU) for a R & D programme. Secondly, the Agence Nationale pour la Valorisation de la Recherche (ANVAR) which funded development programmes for 4 MF (644 kECU) in 1981. Thirdly, the Ministry of Agriculture contributed 1 MF (161 kECU) in 1981 to agricultural demonstration projects. Fourthly, the Association Nationale pour la Récupération et l'Elimination des Déchets (ANRED) and the Ministry of Environment have limited funds for projects involving environmental protection such as landfills. In fifth place are the Agences Régionales de Bassin which give subsidies for agro-industrial projects with environmental protection. Finally, some R & D is funded by the Etablissements Publics Régionaux and the Crédit Agricole bank.

d. Federal Republic of Germany

The Federal Ministry of Food, Agriculture and Forestry and the Regional Agricultural, Economic and Environmental Ministries provide funds to owners of agricultural biogas plants for investment costs in order to introduce better and newer technologies into practice (pilot and demonstration projects with considerable financial contributions by the owner).

The Deutsche Forschungsgemeinschaft (DFG) or German Research Association assists individual scientists and scientific institutions for basic research.

The Federal Ministry for Research and Technology and the Regional (Länder) Ministries for Economic and for Environmental Affairs give financial support to research institutes (which bear no cost themselves), and firms (bearing at least 50 % of the cost themselves) for research and development of biogas projects.

e. Greece

Greek farmers willing to incorporate a biogas plant into their waste treatment system for energy saving and environmental protection can apply to the Ministry of Coordination, submitting the necessary information and the design. Depending on the regional location of the methane digester, from 10 % up to 50 % of the investment costs are reimbursed. These financial aids are foreseen by a new law no. 1262/1982 concerning a large number of industrial activities and including livestock and poultry farms. Biogas plants are not specifically mentioned in this law. To date, no biogas plant which is financially supported by the law is on stream but several applications have already been submitted to the Ministry of Coordination.

Grants for applied research on biomethanation are provided by the Ministry of Agriculture, by the Commission of the European Communities, by the Agricultural Bank of Greece, the latter only for cooperative industrial firms or agricultural farms. In some cases, participation of the owner in the financing of pilot and full-scale biogas plants has been practised.

f. Ireland

The source of funding for the existing biogas plants in Ireland varies greatly. In the industrial sector, pilot-plant installation costs have, to date, been borne by the industries concerned. In the agricultural sector, funding has come from external sources such as the Commission of the European Communities, from commercial interests as in the case of the pig slurry project funded partially by Bailieboro Co-operative, from the Industrial Development Authority, and from the Agricultural Institute as part of its research activities. The full-scale anaerobic filter unit in the Bandon piggery in County Cork is funded largely by the farmer with the assistance of an Energy Savings Demonstration grant from the Department of Energy.

However, there is, at present, no general financial backing for methane digestion projects and no grant scheme which would assist farmers willing to install such a unit.

g. Italy

There are three possible sources of subsidies in Italy. First, essentially in northern and central Italy, Breeder Associations and Chambers of Commerce have programmes for the development of anaerobic treatment of agricultural and industrial wastes. Depending on the cost of the methane digester, the size of farms or industry and local political situations, owners can receive subsidies ranging from 30 to 70 % of the biogas plant cost.

Two other possibilities exist only for activities in the provinces of southern Italy and those in the area of Latina and Frosinone and parts of the area of Rome, Rieti and Ascoli Piceno. There is an ordinary programme of subsidies for all the new activities in these provinces, drawn up by the Cassa del Mezzogiorno. It is possible to obtain without return a sum of money equivalent to one-third of the biogas plant cost, and to borrow another one-third at a reduced interest rate.

A special programme for energy supply in the same provinces was recently approved by the Commission of the European Communities (7 October 1980). A total budget of 9 885 million LIT (7.5 MECU) is granted for the construction of biogas plants. This covers 50 % of the investment cost of the biogas plants and 70 % of the cost for the diffusion of the results.

h. The Netherlands

In The Netherlands, biogas plants are financed by the farmers or industries themselves. However, governmental subsidies on biogas plants altogether amount to approximately 35-40 % of the investment cost. These subsidies are granted as a general subsidy on investment costs (for 8 to 14 % of the cost) as a subsidy on small-scale investments (for maximum 5 % of the cost), as additional subsidy on investments which are beneficial for environmental control (for a maximum of 15 % of the cost). The latter subsidy only applies to part of the biogas plant (digester, storage pits). An additional subsidy on energy producing investments (for maximum 20 % of the cost) only applies to part of the biogas plant (digester, motor generator set). The total subsidy never exceeds 40 % of the total investment cost.

Apart from the governmental subsidies mentioned above, some biogas plants on farms are eligible for grants to cover 25 % of the investment cost (and another 25 % is lent at a low interest rate) within the framework of a biogas demonstration programme set up by the Dutch Ministry of Economic Affairs. All other digesters built to date are granted 10 000 HFL (3 375 ECU) from a fund managed by The Netherlands Institute of Agricultural Engineering (IMAG) regardless of the digester size, investment cost, etc.

i. United Kingdom

Subsidies for farmers wishing to install a biogas plant are available through the Ministry of Agriculture, Fisheries and Food (MAFF). Through the MAFF, the AHGS (Agriculture and Horticulture Grant Scheme) allows a grant of 22 1/2 % on new cattle and dairy developments. If a digester is part of a complete new development, it is eligible, but certain pieces of machinery may be exempt. If the digester is being installed as an improvement to an existing system, only the tank and not the machinery is eligible. There are maximum and minimum limits of expenditure for eligible projects which change depending upon exchange rates, etc. Developments involving pigs have a much narrower range of limits than cattle, and poultry developments are excluded from the AHGS.

Subsidies for industries are available through the Department of Energy and the Department of Industry. The Department of Energy operates an Energy Saving Loan Scheme for industry. This applies to projects costing not less than UKL 10 000 (17.6 kECU) and is available for such things as the installation or modification of plant so as to enable waste organic materials to be used as fuel. A regional development grant under the Industry Act 1972 could perhaps be applied for. The grant would only be given if the biogas plant forms part of a recognized development in specific regions (North of England, Scotland, Wales, Cornwall). The Department of Industry also has a certain amount of money for aiding the development of pieces of equipment. The criterion here is that the idea should have got beyond the research stage. Known as the Product and Process Development Scheme, the aim is to encourage the development of new products or processes with the emphasis on mechanical and electrical engineering. Assistance normally takes the form of a 25 % grant on qualifying costs or a shared contract in which the Government will provide up to 50 % of qualifying costs in exchange for a levy on commercial sales.

j. Switzerland

In Switzerland, the financial support for biogas plants on farms is not very important. Only 5 biogas plants out of 100 have received subsidies in 1982 and not more than 25 % of the investment cost of the biogas installations is paid for in that way. Recently, this financial support has been improved and farmers are allowed to submit their projects to install methane digesters to the Cantonal Authorities. If approved, one half of the financial support is paid by the Cantonal Authorities, the other half is paid by the Central Government. The exact amount of the financial support for alternative sources of energy is not yet precisely identified, and a survey is being conducted by ASETA (Swiss Association for Technical Equipment in Agriculture).

CHAPTER IV. BOTTLENECKS IN THE IMPLEMENTATION OF BIOMETHANATION

Bottlenecks in the implementation of biomethanation in the European Economic Community and Switzerland were identified first by national experience as reflected in national papers written by the experts and secondly, by a scrutiny of the completed survey. These two sources are considered separately below.

IV.1. Bottlenecks identified by national experience

IV.1.1. Agricultural methane digesters

Many farmers have expressed an interest in biomethanation. However, many factors deter farmers from actually investing in a biogas plant. These factors are listed below in order of decreasing importance.

1. Methane digesters in practice exhibit an apparent lack of reliability which has been emphasized in the press and has led to adverse comments on biomethanation.

2. Turn-key methane digesters are still relatively expensive. One of the reasons for this is that the size of the methane digester is frequently determined by the quantity of feedstock available and not by the energy needs of the farm. On the other hand, to lower the costs, potential users put in a large amount of relatively inexperienced labour. The resulting biogas plant usually exhibits a lower efficiency.

3. In some countries such as Denmark and Ireland, the whole agricultural area has economic problems. Moreover in these two countries, the interest rate for money loans is very high (up to 18 %).

4. In some countries such as France, the Federal Republic of Germany and Belgium, the feedstocks available in the largest quantities are those for which a good methane digestion system does not yet exist, e.g. bedding manure.

5. Pollution control is often more loosely enforced for farm wastes than for sewage sludges or industrial wastes, with the exception of Italy and Greece.

6. Conventional fuels of fossil origin remain available and convenient, and their prices have recently fallen in some countries.

7. It sometimes appears difficult to integrate a methane digestion system, i.e. the methane digester and its ancillary equipment, in existing farms. On the other hand, the farms do not always possess an adequate structure to meet their energy requirements.

8. There is a lack of information on the economic benefits of agricultural methane digesters.

9. Not all of the biogas produced can be used easily. Thus, the benefit recovered as energy in the form of biogas is low, and this increases the effective unit cost of the energy produced.

IV.1.2. Industrial methane digesters

Although the prospects of industrial biomethanation seem promising, specific factors still restrain its development.

1. Aerobic biological treatment of wastewaters remains an actively competing alternative to biomethanation, because it has been used and demonstrated for many years.
2. Some recent, highly productive technologies have few working examples or have still to be demonstrated at full-scale level.
3. Even if legislation for water and air pollution control does exist, it is poorly enforced in some countries such as Belgium so that industries are not under stress to treat their wastewaters.

IV.2. Problems of engineering and operation of biogas plants highlighted by the survey

The survey produced on the one hand 96 detailed answers on problems encountered in the practice of biomethanation. On the other hand, in the Federal Republic of Germany, the evaluation of the problems was deduced from the results of an inquiry conducted by Baader and Kloss. A Swiss report, written by Wellinger, relates the problems encountered during the monitoring of 7 Swiss biogas plants on farms during the years 1979-1980.

The problems elicited were classified and the most relevant problems are presented below. Two grades of problems are considered :

1. A problem is called major when it results in the temporary shutdown of the biogas plant or in the souring of the methane digestion liquor.
2. A problem is designated as minor when, although it does not result in a major disruption of the operation of the biogas plant, it is nevertheless costly in terms of component replacement and/or of man-hours involved in maintenance or servicing or is potentially hazardous.

As a general comment, it appears that each biogas plant has suffered during its lifetime an average of 2 major and 4 minor problems.

IV.2.1. Problems with the engineering and maintenance of biogas plants

a. Loading devices

The failure of the feedstock pumping device appears to be an important problem. It was encountered 10 times as a major problem and 15 times as a minor problem.

There exists also a major problem when feeding highly concentrated substrates. There are as yet no pumps for the feeding of wastes with a dry matter content higher than 15 %.

b. The methane digester itself

Problems concerning the engineering and maintenance of the methane digester itself are presented in Table II.4.1. The most important problem is gas leakage occurring in the methane digester. The origin of these gas leakages lies either in the material used for the construction of the digester or in its basic design. Methane digesters constructed in concrete are the most subject to gas leakages. Five batch methane digesters out of 15 suffer the same problem owing to their moving covers. Two out of seven IRIS industrial methane digesters had leaks through their tarpaulins. Finally, two methane digesters with integrated gasholders presented problems of gas leakage through their floating roofs.

Table II.4.1. Problems with the engineering and maintenance of the methane digester itself

Problem	Major	Minor
1. Gas leakage	13*	14
2. Inefficient mixing device	11	8
3. Failure or inefficient heat transfer in heated digesters	6	5
4. Electrical failures	5	2
5. Problems with excess pressure release devices	2	7
6. Corrosion of digester unit and component parts	1	9
7. Problems with insulating materials	2	2

* Number of cases encountered out of 96 detailed answers produced by the survey.

The second most important problem is the inefficiency of the mixing devices with consequent scum formation and/or settlement in the methane digestion tank. All of the most frequently encountered mixing devices, i.e. liquid recirculation, gas recirculation or mechanical mixing, often present problems even when they are working continuously or when two mixing devices are working together. Thirty-four percent of the liquid recirculation devices, 23 % of the gas recirculation devices, 20 % of the combined liquid and gas recirculation devices and 18 % of the combined liquid recirculation and mechanical mixing devices are inefficient. Both cases of combined gas recirculation and mechanical mixing device proved inefficient. Although, strangely enough, they are not frequently used, the only mixing devices without problems are propellers, combined liquid recirculation and propeller devices, and archimedean screws.

The major reason for an inefficient mixing of the methane digester content is not the type of mixing device but more often the non-adaptation of the size of the mixing device to the size and shape of the methane digester. For example, if we consider mixing by gas recirculation, recirculation through draft tubes should be preferred for tall and narrow methane digesters whereas only gas diffusion from the bottom of the digester should be used for wide and low forms.

Although insulating materials seem to cause few problems in Europe as a whole, Switzerland is a special case. Indeed, in some Swiss methane digesters, too high a process energy demand is usually due to poor insulation of the reactor and the loading pipes. The thickness of the insulation is often insufficient (less than 8 cm), especially during winter conditions. The pipes are often too long and are badly insulated. Moreover, in some cases the insulation material absorbs water rapidly and hence becomes inefficient.

c. Biogas collection, storage and utilization devices

An analysis of the problems concerning biogas collection, storage and utilization devices is given in Table II.4.2. It is evident that electricity generation systems present major problems, both in relation to engine operation and waste heat recovery devices. Many engines have too short a life-span, require too much maintenance and/or repair, and cannot easily be integrated into the energy network of the plant.

The problems associated with direct burning devices appear to be minor, although when firing off unpurified biogas, disturbances caused by deposits in the oven occur frequently. Apart from an increase in maintenance, these deposits also lead to a coating on the surfaces of the heat exchangers which considerably lowers the efficiency of the units.

Table II.4.2. Problems with gas collection, storage and utilization devices

Problem	Major	Minor
1. Collection		
1.1. Water blockage of gas lines	1*	17
2. Storage		
2.1. Compression problems during gas pumping	6	5
2.2. Leakage of storage devices	3	5
3. Direct burning devices		
3.1. Problem with design and construction material of burner	1	12
4. Electricity generation systems		
4.1. Engine problems	12	3
4.2. Difficulties with waste heat recovery systems	9	9

* Number of cases encountered out of 96 detailed answers produced by the survey.

IV.2.2. Problems in the operation of biogas plants

a. Problems with feedstocks, their composition, storage and handling

The seven relevant problems concerning feedstock composition are presented in Table II.4.3. Settlement during storage appears to be a particular problem with pig manure. Groundwater or rainwater entry during storage of the feedstock is particularly dramatic because of its inherent low content of dry matter. Feed variation due to changes in factory operation is often a minor problem in the treatment of industrial wastewaters. Scum formation is characteristic of pig manure. Methane production, compositional changes, losses in volatile organic content are reported during storage and become more important as the storage time increases. The start-up and first-year operation of a methane digester with old manure may give rise to major problems.

Table II.4.3. Problems with feedstocks, composition, storage and handling

Problem	Major	Minor
1. Settlement during storage	8*	8
2. Groundwater or rainwater entry during storage	8	4
3. Feed variation due to changes in factory operation	2	14
4. Scum formation	6	4
5. Methane production during storage	3	5
6. Compositional changes during storage		7
7. Loss in volatile organic content	1	4

* Number of cases encountered out of 96 detailed answers produced by the survey.

b. Loading operation problems

Table II.4.4 summarizes the loading operation problems. The most frequently encountered problem appears to be the blockage of pipelines which is always due to the use of pipes with too small an internal diameter. External heat exchangers appear to cause more difficulties than internal heat exchangers. Precipitation in pipelines as a result of high alkalinity is often encountered with strongly buffered industrial wastewaters.

c. Problems with methane digester operation

These problems are summarized in Table II.4.5. Most of them are, in fact, the consequences of inefficient mixing devices.

Table II.4.4. Problems with loading operation

Problem	Major	Minor
1. Blockage of pipes due to undesired material in the waste	6*	15
2. Operational problems with heating of feed	8	3
3. Precipitation in pipelines as a result of high alkalinty	5	3
4. Build-up of deposit on heat exchanger surface causing ineffective heat transfer	2	5
5. Blockage of heat exchanger	0	7

* Number of cases encountered out of 96 detailed answers produced by the survey.

Table II.4.5. Problems with digester operation

Problem	Major	Minor
1. Settlement in the reactors	6*	9
2. Scum formation	7	5
3. Difficulties with gas separation devices	2	6
4. Blockage of the overflow pipe	2	0
5. Start-up too rapid	2	0

* Number of cases encountered out of 96 detailed answers produced by the survey.

d. Chemical and biochemical problems

These problems are listed in Table II.4.6. The presence of hydrogen sulphide, H_2S, can be an important problem especially where electricity is generated and the biogas is poorly scrubbed or not scrubbed. In Switzerland for example, the major technical problems encountered were all related to the relatively high hydrogen sulphide content in the biogas (from 700 to 4 000 ppm). Often, unsuitable metals such as aluminium or copper, or plastic materials such as soft polyvinyl chloride (PVC) or polystyrol used for the gas lines were severely corroded within a short time. The burners often had to be replaced within half a year when the heat exchangers were made of lead-coated copper. Sulphur problems do not stop after firing the biogas because the sulphuric acid formed from H_2S by oxidative burning is also highly corrosive for the exhaust device.

External buffering of the methane digestion mixed liquor gives rise to problems since it requires large quantities of lime or other buffers. Concentrations of ammonia higher than 5 000 mg l^{-1}, as is sometimes the case with pig manure, greatly reduce the biogas productivity. Antibiotics such as monensin or emtryl can give rise to major problems. Sulphide toxicity or inhibition is often encountered when biomethanizing wastewaters from distilleries or starch or baker's yeast factories.

Table II.4.6. Chemical and biochemical problems

Problem	Major	Minor
1. H_2S problems in gas	3*	16
2. Buffering problems	4	2
3. Toxicity and inhibition		
3.1. Ammonia	2	4
3.2. Antibiotics	2	1
3.3. Sulphide	1	2

* Number of cases encountered out of 96 detailed answers produced by the survey.

Table II.4.7. Effluent handling problems

Problem	Major	Minor
1. Storage	3*	7
2. Disposal problems		
2.1. Nitrogen removal	3	3
2.2. Aerobic final treatment	2	3

* Number of cases encountered out of 96 detailed answers produced by the survey.

e. Effluent handling problems

The problems encountered in the handling and treatment of the effluent are presented in Table II.4.7. The ultimate disposal is a major problem with large piggeries, as in Italy.

IV.2.3. Problems in the monitoring of biogas plants

a. Problems with the monitoring itself

The survey elicited three types of monitoring problems which are summarized in Table II.4.8. Gas metering appears to be one of the most important problems highlighted by the survey. Problems with temperature and pH probes are much less important.

b. Lack of data on performance and energy balance

The relative availability of data concerning performance, energy balance or profitability is shown in Table II.4.9. Monitoring does exist but needs coordination and a common language. Information on the performances of methane digesters (gas productivity, yields, depollution...) is available to some extent. Data on energy balance or profitability, even for monitored methane digesters, is notably absent in most instances.

Table II.4.8. Problems with monitoring

Problem	Major	Minor
1. Gas metering devices	19*	10
2. Temperature probes	3	8
3. pH probes	3	7

* Number of cases encountered out of 96 detailed answers produced by the survey

Table II.4.9. Availability of data regarding monitoring of the methane digesters surveyed

Data regarding	Information available	Part of the information available	No information	No gas use	Too early for the information
Performances	53 % (1)	14 %	11 %		22 %
Energy balance	10 %	10 %	50 %	6 %	23 %
Profitability	14 %	2 %	53 %	9 %	24 %
	Exists (2)	Minimal (3)	None	No answer	Too early
Monitoring	68 %	5 %	17 %	5 %	5 %

---

(1) % of the 155 biogas plants for which a complete data sheet was obtained. Each horizontal line sums up to 100 %.

(2) A specialized laboratory or other scientific service monitors the methane digester.

(3) A non-specialized service monitors the methane digester in a routine manner.

CHAPTER V. THE WAY AHEAD : TECHNICAL IMPROVEMENTS FROM PRACTICE AND R & D EFFORTS

As problems and bottlenecks appeared, solutions were sought and research was initiated in a somewhat disorganized fashion. The scattered information elicited by the survey was organized under the main headings of the present chapter.

V.1. Technical improvement through practical experience

Practical experience reported in the individual inquiries for most European Community States and in a written summary for Switzerland (Wellinger, 1982) has indicated a number of preliminary solutions to technical problems which are summarized here.

V.1.1. Practical improvement of mixing problems

It has been seen that mixing devices are often inefficient, the consequence being scum formation and/or settlement in the methane digester. The following improvements of mixing devices originate from practical experience :

- longer working periods of the intermittent mixing device,
- addition of an auxiliary mixing device, often gas recirculation or a high speed propeller, although this requires a higher energy input,
- installation at the centre of the methane digester of a heating-gas mixing device as used in sewage sludge methane digesters, but this applies only in the case of liquid manures.

V.1.2. Practical reduction of heat losses

Although heat loss did not appear as a relevant problem in the survey, it was often encountered during the monitoring of the Swiss agricultural biogas plants in 1979-80. Practical solutions for avoiding heat losses are mainly based on improved insulation :

- the thickness of the insulation layer must be at least 12 cm,
- the insulation material must be protected against moisture,
- in- and outflow pipes as well as heating pipes must be well insulated.

According to the Swiss experience, extruded polystyrene seems to be one of the best insulation materials.

As well as these practical solutions based on improved insulation, heat losses can also be reduced by minimizing the length of heating pipes.

V.1.3. Practical reduction of energy input

The energy demand for heating the influent and for maintaining the temperature of the methane digester can be reduced to some extent in the following ways :

- a heat exchanger can be installed between influent and effluent of the methane digester. Unfortunately, the efficiency of such passive exchange units is very low, particularly for slurries, and blockages occur too frequently;
- methane digestion can be run at ambient temperature or below. But, under these psychrophilic conditions, the start-up of the methane digester takes a long time (up to one year) and the biogas productivity is reduced to as low as 55 % of the value encountered under mesophilic conditions;
- pre-aeration of the manure promotes aerobic biological activity which can raise the temperature of the manure to 20 °C. A Swiss methane digestion system called "Roxistall" is based on this principle. The methane digester with an integrated aerator in the centre is located underneath the round stable;
  - by preventing heat losses of the fresh feedstock (insulation of the storage tanks, pipes, gravity lines...);
  - by avoiding dilution of the feedstock.

Two recommendations arising from practice are aimed at reducing the mechanical energy demand :

- mixing can be generated by using the build-up gas pressure;
- the methane digester can be partly buried in the ground so that loading can occur by gravity.

V.1.4. Practical improvement of biogas handling and uses

Because of the presence of H_2S and other corrosive sulphur-containing compounds in the biogas, the choice of an appropriate material for the gas pipes is of great importance. Among the polymeric materials, polyethylene seems to be the most suitable because of its strength and transparency. In addition scrubbing of the biogas is recommended. Furthermore, precautions must be taken to ensure that water is removed from the pipes.

V.1.5. Practical improvement of the process itself

Some rules have been elaborated in practice with a view to improving the process, especially in the case of industrial wastewaters. These rules are summarized in Table II.5.1.

V.2. Implementation of biomethanation through research and development programmes

In recent years, the number of R & D programmes in biomethanation has grown significantly in all the Member States of the European Community and Switzerland and within the activities of the Commission of the European Communities.

Table II.5.1. Rules derived from practice to improve the biomethanation process

Type of wastewater	System	Modification(s)
- Starch factory	UASB	- Enlargement of the buffer tank
- Potato industry	UASB	- Preacidification
- Pectine industry	Anaerobic contact	- Improved sedimentation tank installed
- Different types of industrial wastewaters (baker's yeast, dairy, cider...)	Anaerobic contact	- Patented settling tanks to avoid washout of biomass
- Wastewater with high concentration of sulphates	Anaerobic contact	- Patented sulphur scrubbing device
- Distillery wastewater	UASB	- Lamellar decantation between the two steps (fermentative acidification and methanogenesis) with recycling of active biomass - Slow mixing device in the first step - Lamellar separation in the decanting step inside the methane digester.

Although bottlenecks for the implementation of biomethanation still remain, as was seen in Chapter IV, the existing R & D programmes are already tackling some of them.

V.2.1. Existing R & D programmes in the European Community Member States and Switzerland

Many R & D programmes on biomethanation in Europe are funded by national governments, public institutions, industries, and also by the Commission of the European Communities. The various tendencies of these R & D programmes are summarized below. The detailed R & D programmes going on in each country of the European Community and Switzerland can be found in Appendix II.A.

a. Basic research

Basic research on biomethanation is essentially devoted to the biochemistry and the microbiology of methanogenesis, particularly the study of methanogenic bacteria. The genetics of bacteria has received less attention. This basic research is presently going on in five countries of Europe : France, Germany, Italy, The Netherlands and the United Kingdom.

b. Applied research

Optimization of the anaerobic digestion of several types of feedstocks is the main topic of applied research on biomethanation. It includes both the optimization of the biomethanation systems themselves (engineering) and optimization of the running conditions of biomethanation (process). In this way, better adapted biomethanation systems i.e. systems with accumulation of biomass, are emerging for the following wastes :

- agricultural wastes : this includes the wastes produced on farms, crop residues and also tropical wastes. All countries covered by the survey are involved in this research;
- industrial wastes : this includes wastewaters as well as solid wastes. Again all countries covered by the survey are active in this field,
- domestic refuse : research in this field is devoted both to operation of test-sites on landfills and to biomethanation of the refuse in reactors. Only Greece and Ireland are not yet involved in this research;
- energy crops : to date, research limited to biomethanation of algae has been undertaken in Belgium, Germany, Italy and France.

c. Demonstration research and monitoring

Demonstration and monitoring of some full-scale biogas plants is presently going on in almost all countries covered by the survey. So far, most of these demonstration programmes have concerned biogas plants on farms.

d. Research on gas utilization

This part of the research concerns essentially the utilization of biogas in engines. It is going on in four countries : Denmark, Germany, the United Kingdom and Switzerland.

e. Research on effluent uses

The fertilizer value of digested effluents is at present studied only in Denmark, Germany and Switzerland.

f. Economics

The economic evaluation of biogas plants is being undertaken in four countries : Belgium, Denmark, France and Germany.

g. Modelling

Finally, modelling of the process of biomethanation is being studied in two countries : Belgium and the United Kingdom.

V.2.2. R & D activities within the programme of the Commission of the European Communities

a. Financial support for the demonstration of biogas projects by the Directorate-General for Energy

On 1 June 1978 the Council of the European Communities adopted a regulation on the granting of financial support for projects to exploit alternative energy sources. For the sector of exploitation of solar energy, which includes biomethanation, a total of 22.5 million European Currency Units (ECU) has been fixed for the five-year programme. The Commission decides whether to grant or refuse support for projects after consulting an Advisory Committee on the management of the projects concerned, composed of representatives of the Member States. Support does not in general exceed 40 % of the total cost of the project. The decision to support has already been taken for four biogas projects. These are detailed in Table II.5.2.

Table II.5.2. Biogas projects supported

Object of demonstration	Company	Level of funding
1) Methane production (NL)	IMAG (Dutch Agriculture Institute)	HFL 122 400,- (ECU 43 922,-)
2) Revninge "Biogas System" (DK)	Revninge Energy Co-operative (REC)	DKR 3 206 000,- (ECU 408 753,-)
3) Development of full-scale biogas plants (B)	AIDR(1)	BFR 12 012 000,- (ECU 297 101,-)
4) The "Energy Farm" (I)	SES(2) Italy	LIT 134 000 000,- (ECU 124 292,-)

(1) International Association for Rural Development;
(2) Sistemi Energia Sud (Soft Energy Systems).

Galway (Ireland) 1982. 3.5 m^3 pilot-scale anaerobic filter digester on pig manure.

Lamezia-Terme (Italy) 1982. 1 m^3 pilot-scale completely mixed digester on algae. In front, view of the basins of algae growing.

b. Research in the field of biomethanation coordinated by the Directorate-General for Science, Research and Development

- Within the Research Programme on Primary and Secondary Wastes

Within the framework of the Indirect Actions on Environmental Protection and Recycling of Industrial and Urban Wastes, 28 contracts have already been awarded. They can be classified as follows :

- four contracts on specific studies of enzymatic processes. Celluloses, cellobioses and wastes containing cellulose can be treated in a minor way by methane digestion;
- sixteen contracts on the processes themselves. Wastes of agricultural and industrial origin can be treated in a major way by methane digestion;
- nine contracts on pilot-scale experiments. Fibrous residues from agricultural wastes, industrial wastes, domestic refuse are treated by anaerobic digestion;
- five contracts on "farm scale" and "field experiments". Agricultural wastes and domestic refuse are here mainly considered for methane digestion;
- one contract on the "industrial" pilot experiment on the mixed treatment of effluents from pulp and paper industries by methane digestion as well as by fermentation and hydrolysis;
- one contract on the definition of methods for controlling and managing a methane digester;
- one contract on the survey of corrosion of digesters, fermenters and storage tanks.

- Within the Programme in the field of Solar Energy

Within the Programme of Energy from Biomass, the Commission of the European Communities has sponsored several research projects in the field of biomethanation :

- on mariculture and aquatic wastes (4 contracts) : feasibility of digestion, optimization of the process, pilot demonstration at real site;
- on agricultural wastes and energy crops (3 contracts) : optimization of the processes of biomethanation;
- of agro-food wastewaters (1 contract) : optimization of the process of biomethanation.

c. Financial support of biogas projects in developing countries by the Directorate-General for Development

The Commission of the European Communities has financed a number of projects in new and renewable energy in some countries of Africa, the Carribean and the Pacific (ACP-countries), as well as in some non-associated countries which have agreements with the Community. Some biogas projects have already been developed in Africa and India since 1980 (see Table II.5.3). The objective of this financial support is to permit the integration of simple methane digesters in these countries, well adapted to local conditions and acceptable by local people. Another project consists in the construction of a Regional Centre of Solar Energy in Upper Volta which includes a laboratory of biomethanation (contribution of the European Community = 3 million ECU).

The objectives of this Centre are to study, apply and demonstrate different technologies of energy production, including public awareness activities and training of local people. Other biogas projects are projected in Togo (oil-mill waste) and in Thailand (potentiality on farm). These last two projects are supported but not funded by the European Community.

Table II.5.3. Biogas projects in developing countries, financially supported by the European Community

Country	Non-Governmental Organisation (NGO)	Title and site of project	Total cost of project, ECU(1) (EC(2) contribution)
Rwanda	AIDR(3)	Production of methane gas from organic waste in various places	467 489 (289 843)
Burundi	AIDR	Production of methane gas from organic waste in Cibitoke, Kibumbu and Kiriri	264 792 (172 115)
Upper Volta	AIDR	Reduction of firewood consumption via the production of methane gas from organic fertilizer in the rural areas around Ouagadougou, Kaya, Kongoussi and Goundy	203 951 (144 805)
India	Deutsche Welthungerhilfe (DWHH)	Promoting cattle feed production and building biogas plants at Narayangoan in the State of Maharashtra	82 146 (40 877)

(1) ECU : European Currency Unit;
(2) EC : European Community;
(3) International Association for Rural Development.

d. Financial support for biogas plants in Southern Italy by the Regional European Fund of Development

A special programme for energy supply in the provinces of southern Italy was approved on 7 October 1980 by the Commission of the European Communities within the framework of the Regional European Fund of Development. Within this programme, subsidies are assigned to the construction of several biogas plants. The total budget of 9 885 million LIT (7 460 000 ECU) covers 50 % of the costs for the construction of the plants and 70 %

of the costs for the diffusion of the results. The 32 breeding units (18 on cow manure, 12 on pig manure, 2 on poultry manure) concerned by this project are distributed by province as follows : Abruzzi 7; Basilicata 4; Calabria 2; Campania 5; Lazio 3; Marche 2; Molise 5; Puglia 1; Sicilia 1; Sardegna 2.

V.3. Recommendations for future R & D

In order to overcome the bottlenecks elicited by the survey and not yet solved by the existing research programmes, seven recommendations can be put forward as suggestions for future R & D.

V.3.1. Network monitoring of selected biogas plants in Europe

An absolute priority must be given to the monitoring of a number of biogas installations which would serve as European demonstration plants. To this end, a European network should be drawn up on much the same principles and in the same way as the present CEC working group on "Biogas Plants in Europe". National monitoring programmes are known to exist in Denmark, The Netherlands and Switzerland; others are starting in France, the United Kingdom, Belgium and Germany. The European network should be set up in such a way as to complement these national efforts and provide them with additional resources. The data collected should concern the biological as well as the engineering aspects of the biogas plants and include energy balances and cost analyses. These data should, first, be largely diffused among scientists and engineers and, secondly, lead to a substantial popularization of the biomethanation process. In this way, a confident market for biogas plants will be promoted.

In order to achieve such a task, a common language needs to be used within the network. Therefore, within this European network, the standardization of methods, parameters, units and symbols needs to be promoted. A compendium of analytical and measurement methods, used within and outside the European Community should be the first objective. Inter-laboratory checking of questionable methods could be the second aim. These efforts could lead to the publication of a manual of acceptable standard methods.

V.3.2. Basic research

There is a need for a better understanding of the basic processes on which biomethanation relies. The research should aim at improving yields and rates in existing processes, at including new biomasses (wastes, residues or energy crops) as possible substrates for biomethanation and at obtaining a larger variety of energy-yielding molecules (e.g. octane boosters).

Emphasis should be given to fields of microbial physiology and microbial ecology including micro-ecology, i.e. the ecology of carrier-bound mixed cultures. Although the application of genetic engineering techniques should be considered, a cautious approach is recommended in view of the present lack of basic knowledge of the microorganisms which could or should undergo DNA manipulation or of the laws by which a given species is accepted or rejected within a microbial community. The basic research should lead ultimately to cheap bulk fermentation processes, by which a large variety of biomasses could be bioconverted to a number of energy-rich chemicals, thereby contributing in its way and for its part to the long term stabilization of the European Community and world chemical resource base.

V.3.3. Improvement of the engineering aspects of biomethanation

Existing biogas plants still suffer two major drawbacks. The reliability with time of the methane digester itself as well as the ancillary equipment is neither well known nor guaranteed. Emphasis should be placed on R & D projects aimed at the comparative evaluation of the reliability with time of a large range of standard or newly developed equipment. Such an evaluation could lead to the establishment of a procedure for approval of the equipment at the European Community level.

Secondly, there is a need for the development of cheap, reliable methane digesters in order to make the present market more profitable but also to open this market quite substantially e.g. by making methane digesters attractive to a larger number of smaller farms.

V.3.4. Basic technology for second-generation methane digesters

A number of newly developed methane digestion systems are being assayed on a small scale. The preliminary results look very promising. These second-generation methane digesters involve the immobilization of mixed cultures on various carriers and the use of these carrier-attached active biomasses in fixed film or fluidized bed reactors.

There is a need for the promotion of R, D & D projects aimed at improving the bases for such technologies. Typical aspects include mass transfer, hydraulic rates, microbial growth within these reactors, reaction rates, etc.

Mathematical modelling based on progress in the above aspects will then permit appropriate simulation of the processes with a view to optimization and management.

V.3.5. New or improved end uses

Upgrading the side products of the biomethanation process will give an added value to these products and provide a valuable incentive. Therefore, more attention should be given to the potential uses of liquid and solid effluents from biogas plants and the possibility of sale after treatment.

V.3.6. Integration of biogas plants within their surroundings

A substantial improvement in the economics of methane digestion systems can be derived from an improved integration of the methane digestion system within its surroundings, whether farm or agro-industry. The energy loop consisting of a well integrated and managed biogas plant can lead to autonomy in energy of the farm, a self-sufficient goal by itself.

R, D & D projects should be promoted which lead to a better integration of methane digestion systems in the overall planning of the plant on the farm. These projects should include the modelling aspects of the problem and examine, among others, the feasibility of collective biogas plants.

V.3.7. Safety, legislation and standardization

Attention must be paid to efforts aimed at bringing existing laws and codes of practice on safety into conformity with the potential of biomethanation and also developing new safety standards in step with progress. Such efforts have already begun in Italy (Anon. a), in the United Kingdom (Anon. b) and in Switzerland (Anon., 1979). Sanitation aspects i.e. the possibility of using biomethanation as a pathogen killer should be included in these efforts.

The Working Group on Biogas Plants in Europe, at its June 1982 meeting in Braunschweig, agreed by consensus upon the preceding recommendations which are intended to serve as guidelines for a future policy of R, D & D promotion by the Commission of the European Communities.

CHAPTER VI. SCENARIO FOR THE FUTURE

A scenario for the implementation of biomethanation may be constructed on the following lines :

- biomethanation on farms, large and small scale,
- biomethanation in industry, particularly the agro-industries,
- biomethanation in developing countries,
- biomethanation of urban refuse.

This scenario will take into account, first, the possible share of biomethanation among alternative energy sources, and secondly, the place of biomethanation within developing biotechnologies.

VI.1. Possible share of biofuels among alternative energy sources

The possible share of renewable biofuels obtained from biomasses through biotechnology in the current energy demand in industrialized countries may be considered from either a relative or an absolute point of view. The relative potential share of energy from biomass through biotechnology is low, around 7 % as estimated today, but the possible absolute amount of energy from bioprocesses is large. The potential of all forms of biofuels for Europe has been estimated at 75 Mt oil equivalent per year (Nyns *et al*., 1982).

Among the biotechnologies which convert biomass to energy, the biomethanation route offers the advantage of a mature technology, economically attractive by itself in the immediate and short-term future (0-5 years) and by a number of fringe benefits from environmental protection to nutrient recycle in nature.

VI.2. Biomethanation on farms

In the early part of the short-term range (0-5 years), methane digesters will be put on the market by a large number of small firms, some of them being subsidiaries of large companies, diversified in various aspects of biotechnology. These methane digesters will be of the once-through, completely mixed type with continuous or intermittent loading of slurries, essentially animal manure. Because they will be constructed in small series, the price of digesters will diminish, and hence their size at the lower level at which they become economic will match the feedstock available from 60 livestock units. As farms are rebuilt or extended, more and more methane digestion systems will be installed and they will be progressively better integrated into farm management. Furthermore, simplified and robust methane digesters will become available as a "do-it-yourself" technology and progressively extend the market to farms of all sizes.

Financial incentives will be needed initially to promote the market, mainly in the form of public subsidies and low interest rates. Society will be compensated by the added value of environmental control (odour and BOD) of the process. At first, potential buyers will come from farms with large energy demands, say more than 2 500 ECU per year. But as the market develops, the principal attraction will be the self-sufficiency in energy of farms equipped with methane digesters. The assurance of a long-term, autonomous energy supply, often greater than the farm energy demand, will promote overall farming through further investment in energy-demanding processes such as animal nurseries, hay-drying etc.

In the latter part of the mid-term range, progress brought about by R & D projects will open the market to other agricultural wastes as feedstocks for biomethanation, namely solid wastes among which manure with bedding. This will apply not only to industrially constructed larger scale methane digesters (60 m^3 working volume) but also to "do-it-yourself" smaller scale digesters.

Cooperative biomethanation, i.e. the operation of a methane digester by a group of farmers, should be mentioned but its prospects as a "biomethanation on farm" process cannot be evaluated at present.

VI.3. Biomethanation in industry

The technology of biomethanation is already the most advanced in the field of agro-industrial wastewater treatment. In this field biomethanation is a serious competitor for the more classical aerobic biological unit operation in the combined biological and physico-chemical overall treatment process, essentially because a former energy-requiring technology is replaced by an energy-yielding technology.

The biomethanation of agro-industrial wastewaters will take place in a one- or, more often, two-step methane digestion system. The first step, fermentation, will be closely integrated in the agro-industry, possibly as an equalizing storage step. The second step, methanogenesis, will take place in a reactor with a high loading rate, a high methane productivity and a low mean hydraulic detention time. For this purpose, the active biomass, i.e. the microbial bioconverting community, will be either trapped in the reactor or immobilized on a fixed film or on fluidized microcarriers.

It is anticipated that in the earlier part of the short range future (0-5 years), biomethanation will become a classical biological step in agro-industrial wastewater treatment, displacing thereby the aerobic biological step to a mere finishing operation whenever ammonia nitrogen and/or effluent BOD must be reduced to low normative values.

As the process is progressively made to work at lower temperatures, the technology of biomethanation will be adopted increasingly for wastewaters of lower strengths, expressed e.g. as BOD. Hence, from agro-industrial wastewaters, biomethanation will be extended, in the latter part of the mid-range future, to the more dilute domestic or municipal wastewaters.

Similarly, as the process is applied to an even wider range of feedstocks, biomethanation will be extended to a greater variety of industrial wastewaters of non-agricultural origin. Also, from now on, solid industrial wastes will be considered for biomethanation either as a mixture with the wastewater or in a separate methane digestion system. In the latter case, the methane digester will usually be integrated as a unit operation in a global treatment technology including besides the biological step, physico-chemical unit operations. For example, one could imagine that a semi-solid waste would first be pressed; the liquid part would then be further biomethanized in an upflow methane digester, and the liquid effluent discharged. The solid part could be incinerated or, if dried (using the biogas), utilized as animal feed or fertilizer.

Progressively better integrated within industry, biomethanation will become an inherent part of an overall production technology, not only treating wastes or wastewaters to be disposed of or recycled, but employed as an energy loop to make the process less energy-demanding.

VI.4. Biomethanation in the developing countries

Within the next ten years, biomethanation will undergo extensive development in the developing countries.

It has been recognized at the United Nation level (Anon, 1981a) that biomethanation could become, along with mini-hydro power and mini-solar energy converters, one of the major energy sources in developing rural areas. Family-size and small-scale community methane digesters of the do-it-yourself type will contribute to the local energy supply and autonomy. Built with local materials, these methane digesters are expected to spread more through international cooperation than through industrial promotion.

Also at the United Nation level, it has been noted (Anon, 1981b) that "programmes of family-size methane digesters have already been successful, e.g. in China, where designs have been adapted to local conditions. Elsewhere, some designs have been out of financial reach of populations. Poor management and maintenance of biomethanation systems contributed to the failure of many such units. The importance was stressed of developing better designs, of availability of proper management and adequately trained maintenance personnel appropriate to the economic, social, cultural and institutional conditions of the rural populations".

Furthermore, it should be remarked here that in both industrialized and developing countries, agro-industries have hitherto been designed solely for the production of one (or a few) marketable products : e.g. a brewery produces only beer, an alcohol fermentation plant produces only ethanol. With the increasing concern for environmental protection and for saving energy paid for with foreign currencies, agro-industries in developing countries will incorporate biomethanation for this dual purpose. The know-how developed in Europe, as described in the preceding section, will have to be transferred to developing countries; this will imply opening the industrial market for methane digestion systems.

Whereas the development of biomethanation as an energy source is expected in the earlier part of the short-term range, the later part of the mid-range future will see the development of new agro-industries better integrated into their environment and self-sufficient in energy. Biomethanation is expected to play an important role in making these integrated industries more economical. European firms proposing turn-key factories will have to take this into account. For example an alcohol fermentation plant based on amylaceous species such as cassava, will include a methane digestion system for the treatment of wastes, for the agronomic recycling of nutrients on the fields producing the cassava, and as an energy loop which will contribute to make the process energy-autonomous. Other typical examples are the olive oil and palm oil industries, e.g. in Africa.

VI.5. Biogas refineries

Urban refuse has been little used hitherto in biomethanation. This situation is expected to change in the earlier part of the mid-range future owing to the increasing number of landfills being exploited for biogas production over a long term. As there is usually no use for the energy produced at the landfill itself, the biogas produced must be distributed or sold to potential local users.

If we take into consideration, first, that urban refuse must be and already is being collected and transported to a few single point locations; secondly, that landfills will have promoted the idea that urban wastes are potential feedstocks for biomethanation and that the biogas produced thereby can be sold, it may be expected that urban waste discharges will become the starting point of biogas refineries. Very large scale, highly productive methane digestion systems will daily produce large amounts of biogas for industrial or municipal sale and use.

The key to success may be, first, the integration of these energy traps into the existing treatment procedures for domestic waste (shredding, sorting, incineration...) and secondly, the challenge to industry to develop economical, high-performing technologies. Other industrial and/or agricultural wastes will be collected within 10-20 km and biomethanized together with domestic refuse, as is already done with sewage sludge.

In the later part of the mid-range future, biogas refineries will thus appear as a potential complementary source of town gas to be distributed through the existing pipeline network.

VI.6. The place of biomethanation within biotechnologies

The Commission of the European Communities has set up a Programme for the "Forecasting and Assessment of Science and Technology" (FAST) (Anon., 1978) which includes a sub-programme dealing with "Bio-Society" (Cantley, 1980). Within this sub-programme, the German association DECHEMA (Deutsche Gesellschaft für chemisches Apparatewesen eV), on behalf of the European Society for Biotechnology, has evaluated a "Community Strategy for European Biotechnology" (Anon., 1982a, Behrens and Buchholz, 1982).

In this strategy, biomethanation appears as a mature technology generally well served by on-going research trends, as compared with the other biotechnologies. As far as the process of methanogenesis is concerned, the basic research lines presently carried on in the biochemical aspects of biotechnologies, e.g. on the hydrolysis of cellulosic material, are well aligned with the needs for an improved methanogenic process. In contrast, where the microbiological aspects are concerned, the general trend in biotechnology favours genetic engineering whereas biomethanation would benefit more from progress in microbial physiology and ecology, which are somewhat overlooked. The progress being made in enzyme and cell immobilization has direct applications to improved processes of biomethanation such as fixed-film methane digesters.

As regards the technology of biomethanation, the progress being made in the engineering aspects of biotechnology e.g. in mixing, heat and mass transfer, as well as control through modelling also serves the particular requirements of biomethanation.

There is little competition with other biotechnologies for feedstocks, in any event less than in the case of alcohol fermentation. When energy crops are claimed for biomethanation, these crops can belong to non-food species, particularly adapted to grow on land not suitable for food-yielding agriculture. Besides, biomethanation is complementary to these technologies because it can be integrated within them to recycle wastes and residues as energy and other useful end products.

Hence, as a conclusion of the FAST inquiry, biomethanation appears as a generally acceptable and accepted biotechnology, with good strategic prospects both in the field of energy from biomass and in the field of environmental biotechnology.

The foregoing scenario is based on three assumptions : first, that the technical and economic state of the art of biomethanation will be improved; secondly, that the legal policy for the management of the environment is enforced; and thirdly, that the tensions in the energy market at the political and economic levels will persist.

APPENDIX I. HISTORY AND METHODOLOGY OF THE SURVEY

I.1. History of the "Biogas Plants in Europe" project

The idea to survey all biogas plants in the European Community was born in September 1980 during a visit to a farm biogas plant in Duiven (The Netherlands) after a contractors' meeting in Amsterdam (Commission of the European Communities Programme "Solar Energy", project E "Energy from Biomass"). From the very beginning, the general wish was to start monitoring selected biogas plants in the frame of that study. It was, however, decided to focalise the study initially on the collection and scrutiny of available data on existing biogas plants.

It was suggested to set up a Working Group, called at first the "Performance Monitoring Group" (PMG). During its meeting in June 1981 in Copenhagen (Denmark), each member of the Expert Group E (European Commission Programme "Solar Energy", project E "Energy from Biomass") was invited to nominate a candidate representative from his country. This led to the constitution of a Working Group composed of a coordinator, an adviser and national experts. The different members of this group are listed at the beginning of this book under the title "contributors".

The coordinators's tasks are defined in contract no. ESE-R-050-B. The sub-contractor's tasks are defined in contracts nos. ESE-R-051 to ESE-R-058. The cost of the survey is evaluated at 130 000 ECU, VAT excluded, and is entirely financed by the Commission of the European Communities.

In July 1981, the coordinator and the adviser agreed to propose to the Working Group that the survey would be based on inquiries made for each methane digester individually in all Member States of the European Community. Agreement was reached on a draft format of the questionnaire to be submitted to the Working Group.

The Working Group met informally during the Second Symposium on Anaerobic Digestion held in Travemünde in September 1981. The first formal meeting of the Working Group took place in Louvain-la-Neuve on 24 September 1981. It was accepted to base the survey on inquiries made for each methane digester individually. Agreement was reached on the final format of the questionnaire.

The second meeting of the Group took place in Rhenen (near Wageningen, The Netherlands) on 1 and 2 March 1982. At that time, the surveys were well under way for most of the countries. A preliminary statement on the situation of biomethanation was presented by each representative, and the draft of the final report was discussed. It was also decided to add national papers for each country to the survey and to include Switzerland in the study because of its large experience of 108 methane digesters.

The third meeting of the Working Group was held in Braunschweig (F.R. Germany) on 21 and 22 June 1982. At that time, surveys and national papers were almost completed. After a brief review of the results, recommendations for future research and development in the field of biomethanation were agreed upon.

A preliminary report on the status of biogas plants in Europe was submitted to the Commission of the European Communities in December 1982.

The fourth and last meeting of the Working Group was held in Reggio Emilia (Italy) on 28 and 29 March 1983. Results on economic and biological aspects were presented and discussed. A draft table of contents for the present book was proposed, based on the preliminary report, and agreed upon.

I.2. Methodology of the survey

The survey was carried out by the different national experts, each in his own country. Whenever possible, the experts personally visited the biogas plants and questioned the owner and/or the constructor and/or the monitor of the biogas plants. As regards the format of the questionnaire, besides a general information sheet, five different subjects were approached.

1. "Design" of the digestion system :
   - general flowsheet of, and information on the complete digestion system,
   - draft of, and information on the structural and mechanical design of the methane digester.
2. "Process" of the digestion system :
   - feedstocks : type, aspects, eventual treatment, analytical data,
   - influents : aspects, analytical data,
   - type of process,
   - running conditions of the digester,
   - gas : treatment, uses,
   - effluents : treatment, uses, analytical data.
3. Economic aspects :
   - energy balance,
   - costs.
4. General comments :
   - modifications of existing digesters and reasons therefor,
   - motivations,
   - possible developments : available feedstocks, interest of other people...
5. "Problems" encountered :
   - problems with the engineering and maintenance of biogas plants,
   - problems with the operation of biogas plants,
   - problems with the monitoring of biogas plants.

I.3. FORMAT OF THE SURVEY

1. GENERAL INFORMATION

For the following questions, give the name and address

1.1. LOCATION OF THE METHANE DIGESTER

1.2. 1. DESIGNER (consultant) :

2. CONSTRUCTOR :

3. OWNER :

4. SPONSOR OF CONSTRUCTION :

5. FUNDING AUTHORITY FOR CONSTRUCTION :

6. RESEARCH BODY RESPONSIBLE FOR THE FOLLOW-UP :

7. SCIENTISTS PARTICIPATING IN THE RESEARCH :

Give the date and the duration in months or years

1.3. 1. DIGESTER STARTED :

2. DURATION OF OPERATION AT PRESENT :

3. DURATION OF OPERATION INTENDED :

2. DESIGN OF THE METHANE DIGESTION SYSTEM

2.1. GENERAL FLOWSHEET AND INFORMATION OF THE OVERALL SYSTEM

2.1.1. Give a copy of the overall flowsheet of the digestion system

2.1.2. Information on the equipment for storage and treatment before digestion (1)

- Feedstock storage system :
- Feedstock treatment tank :
  - . dimensions (in m) :
  - . volume : m^3
  - . stirring device :

2.1.3. Information on storage and treatment of the gas after digestion

- Gasholder (2)
  - . balloon + trade mark ☐

---

(1) specify
(2) mark the correct answer by crossing the corresponding square

. compressor ☐

. floating gasholder ☐

. other (1)

- Material used for the gasholder (2)

. mild steel ☐

. galvanized iron ☐

. polymeric ☐

. other (1)

- Pressure in the gasholder (1)

. high : mm water gauge or kg/cm^2(3)

. mean : mm water gauge or kg/cm^2(3)

. low : mm water gauge or kg/cm^2(3)

- Treatment of gas

. elimination (2) of . CO_2 ☐

. H_2S ☐

. H_2O ☐

- System used (1)

. CO_2 :

. H_2S :

. H_2O :

. efficiency of the treatment (1)

	before	after
. % H_2S		
. % CO_2		
. % H_2O		

2.1.4. Information on the equipment for the storage and treatment of the effluent after digestion

- Effluent storage system (1)

(1) specify
(2) mark the correct answer by crossing the corresponding square
(3) underline the correct answer

- Effluent treatment tank (1)
  - . dimensions (in m) :
  - . volume : m^3

2.2. DESIGN AND CONSTRUCTION OF THE METHANE DIGESTER ITSELF

2.2.1. Give a copy of the draft of the digester

2.2.2. Characteristics of the digester (1)

- total volume m^3
- mixed liquor volume : m^3
- dimensions height : m
  diameter : m
  other : m

2.2.3. Material used for the digester (2)

- glass ☐
- concrete ☐
- polymeric material ☐
- steel ☐
- other (1)
- internal coating (1) :
- external coating (1) :

2.2.4. Insulation material (2)

- glass wool ☐
- mineral wool ☐
- polymeric insulation ☐
- value of k (1)
- thickness of the insulation (1) :

2.2.5. Heating system (2)

- internal ☐
- external ☐

Heating device (2)

- serpentine ☐

---

(1) specify
(2) mark the correct answer by crossing the corresponding square

- electric heater ☐
- double envelope ☐
- heat exchanger ☐
- other (1)

2.2.6. Stirring system (2)

- gas recirculation ☐
- liquid recirculation ☐
- mechanical ☐
- other (1)

Stirring device (2)

- agitation axle ☐
- motor ☐
- pump ☐

2.2.7. Loading system (2)

- continuous ☐
- intermittent ☐

Loading device

- pump ☐

3. PROCESS OF THE DIGESTION SYSTEM

Remark : whenever possible, answer also Annex I

3.1. FEEDSTOCK

The feedstock is the raw material to be used for the loading of the methane digester as it comes, before any eventual treatment.

3.1.1. Type of Feedstock (3)

- cow manure - pig manure - poultry manure - crop residues - algae - industrial waste - wastewater - refuse.
- other (1)
- mixed feedstocks (1)

---

(1) specify
(2) mark the correct answer by crossing the corresponding answer
(3) underline the correct answer

3.1.2. For biomethanation on farms, answer also Annex II

3.1.3. Aspects of the feedstock (2)

- liquid with fine suspended solids ☐
- semi-solid with rough suspended solids (straw, paper, wood) ☐
- solid ☐

3.2. TREATMENT OF THE FEEDSTOCK (2)

- time storage (1)
- dilution ☐
- grinding ☐
- other (1)

3.3. INFLUENT LOAD

The influent load is the organic load as it enters the methane digester and is obtained from the corresponding feedstock, eventually after a treatment.

- Aspects of the influent (2)
  . liquid with fine suspended solids ☐
  . semi-solid with rough suspended solids (straw, paper, wood) ☐
  . solid ☐

3.4. DESCRIPTION OF THE PROCESS (2)

- Continuous ☐
- Intermittently loaded ☐
- Discontinuous (batch) ☐
- Completely mixed digester ☐
- Plug-flow ☐
- Two-step ☐
- Up / Down - flow with / without support ☐ ☐
- Anaerobic filter ☐
- Fluidized bed ☐
- Anaerobic contact ☐

---

(1) specify
(2) mark the correct answer by crossing the corresponding answer

3.5. RUNNING CONDITIONS OF THE PROCESS (1) (4)

3.5.1. Stirring frequency (1)

- working time : min d^{-1}
- how often per day :

3.5.2. Loading frequency

- working time : min d^{-1}
- how often :

3.5.3. Loading characteristics

- flow rate of the influent : min d^{-1}
- loading rate : kg VS d^{-1} or kg COD d^{-1}
- volumetric load : kg VS m^{-3}_{ML} d^{-1} or kg COD m^{-3}_{ML} d^{-1}

3.5.4. Hydraulic mean residence time (θ) : d
Solid mean residence time (θ_S) : d

3.5.5. Temperature : °C

3.6. PERFORMANCE OF THE DIGESTION (1) (4)

3.6.1. Gas production

- Conditions under which gas is measured (1)
  - . temperature : °C
  - . pressure : kg/cm^2
- Rates of biogas production

  m^3 d^{-1}

  m^3 m^{-3}_{ML} d^{-1}
- Yields of biogas production

  m^3 kg^{-1} VS_o

  m^3 kg^{-1} COD_o

  m^3 kg^{-1} VS_r

  m^3 kg^{-1} COD_r

---

(1) specify
(4) see Annex V : Table of symbols

3.6.2. Gas composition (mean values)

- CH_4 : %
- CO_2 : %

3.6.3. Depollution (see also annex I)

- COD reduction %
- BOD reduction %
- VS reduction %
- TS reduction %

3.7. GAS USES (1)

- Internal use :
- External use :

	% domestic use	% professional use
. hot water		
. heating		
. mechanical power		
. electric power		
. cooking		
. refrigeration		
. lighting		

- Sale
  - . electricity : %
  - . gas : %

3.8. EFFLUENT

3.8.1. Treatment of the effluent (2)

- Decantation ☐
- Filtration ☐
- Other (1)

3.8.2. Uses of the effluent (2)

- Soil conditioner ☐
- Recirculation into the digester ☐

---

(1) specify
(2) mark the correct answer by crossing the corresponding square

- Discharge ☐

- Other (1)

4. ECONOMIC ASPECTS

4.1. ENERGY BALANCE (1)

4.1.1. Total energy required for the biomethanation plant operations PER DAY (= the energy actually used by the boiler or/and the motor)

	MJ	kWh
- Stirring		
- Loading		
- Digester heating		

4.1.2. Equipment used for energy production whenever the produced biogas is used

- Yield of the boiler : %

- Yield of the motor : %

4.1.3. Portion of the biogas produced used for energy production

- In percent : %

- Corresponding to : MJ/day

4.1.4. Net energy recovered per day

	MJ	kWh
- In gas		
- In electricity		
- In warm water		

4.2. COSTS (specify the currency used) (1)

4.2.1. Investment costs

- Total :

- For digester :

4.2.2. Maintenance and operation costs (on yearly basis)

- In money

- In manpower (in man x hour per day) (worker or technician)

- Interest rate for borrowed money

(1) specify

4.2.3. Energy demands of the farm, replaced by biogas (on yearly basis)

- Electricity (kWh) x retail price :
- Warm water production (MJ) x retail price :
- Heating (MJ) x retail price :

:

4.2.4. Energy sold (on yearly basis)

- electricity (kWh) x retail price :
- Biogas (MJ) x retail price :

4.2.5. Effluent sold (on yearly basis)

- Effluent (kg) x retail price :

5. <u>GENERAL COMMENTS</u>

Answer also Annex IV

5.1. EVALUATION OF THE DIGESTER PERFORMANCES (1)

- Is the original design of the digester system satisfactory ?
- Modifications ? (3) Yes - No
  if yes
- When ?
- How ?

5.2. MOTIVATIONS OF THE OWNER(S) ? (2)

- Depollution ? ☐
- Recovery of energy ? ☐
- Other (1)

5.3. ARE THERE POSSIBLE DEVELOPMENTS IN YOUR AREA ? (3)

- Other people interested ? Yes - No
- If yes, specify :
- Other feedstocks available ? Yes - No
- If yes, specify :

---

(1) specify
(2) mark the correct answer by crossing the corresponding square
(3) underline the correct answer

ANNEX I : ANALYTICAL DATA

Answer the following questions whenever the data are available

I. RESULTS

	Influent	Mixed liquor	Effluent	% reduction
- TS (g l^{-1})				
- VS (g l^{-1})				
- COD (mg l^{-1})				
- TOC (g l^{-1})				
- VSS (g l^{-1})				
- pH				
- N_{TK} (g l^{-1})				
- N_{NH4^+} (g l^{-1})				
- VFA (g l^{-1})				
. VA_2				
. VA_3				
. iVA_4				
. VA_4				
. iVA_5				
- Alkalinity (acid mg l^{-1})				
- Organic matter (in % VS)				
. lipids				
. proteins				
. easily hydrolysed				
. polysaccharides				
. hemicellulose				
. cellulose				
. lignin				

Analyses	. Sampling (1) . Pretreatment of the sample	Methods used	
- N_{TK}		1. Method Kjeldahl	☐
		2. Method MacKenzie and Wallace	☐
		3. Other (1) :	
- N_{NH4^+}		1. Titration after distillation at pH 7.4 of NH_3 trapped in boric acid	☐
		2. Method of Chaney and Marbach	☐
		3. Other (1) :	
- VFA		1. Gas liquid chromatography	☐
		2. Other (1) :	
- Lipids		1. Method of Kahane and Mercier	☐
		2. Other (1)	
- Proteins		1. Method of Handing and MacLean	☐
		2. Other (1)	
- Easily hydrolyzed polysaccharides		1. Reducing sugars estimated by Luff-Schoorl method	☐
		2. Other (1) :	
- Cellulose hemicellulose		1. Method of Goering and Van Soest	☐
		2. Method of De Wit	☐
		3. Other (1) :	
- P as phosphate		1. Flame photometry	☐
		2. Other (1) :	
- L(+) Lactate		1. Method of Gutman and Wahlefeld	☐
		2. Other (1) :	
- Formate		Specify the method	
- Ethanol		Specify the method	
- ATP		Specify the method	
- F_{420}		Specify the method	

(1) specify

	Influent	Mixed liquor	Effluent	% reduction
- Mineral matter				
. CO_2 as $CaCO_3$ ($g\ l^{-1}$)				
. Ash ($g\ l^{-1}$)				
. Pi ($g\ l^{-1}$)				
- ATP ($g\ kg^{-1}$)				
- Coenzyme F_{420} ($mol\ l^{-1}$)				
- Formate ($g\ l^{-1}$)				
- Lactate ($g\ l^{-1}$)				
- Ethanol ($g\ l^{-1}$)				

II. <u>ANALYTICAL PROCEDURES</u>

Analyses	. Sampling (1) . Pretreatment of the sample	Methods used	
- TS		1. Drying at 105 °C (12 hours) 2. Other (1) :	☐
- VS		1. Residue between 105 °C and 450 °C (12 hours) 2. Other (1) :	☐
- COD		1. Standard methods : specify the oxidation time 2. Other (1)	☐
- TOC		1. Organic carbon oxidized in CO_2, CO_2 reduced in CH_4 which is measured by flame ionization 2. Other (1) :	☐
- CO_2 as $CaCO_3$		1. Residue between 450 °C and 10 °C x 2,27 (3 hours) 2. Other (1) :	☐
- Ash		1. Residue at 1000 °C (3 hours) 2. Other (1) :	☐

(1) specify

ANNEX II : SUPPLEMENTARY REQUESTS FOR BIOMETHANATION ON FARMS

1. TYPE OF BREEDING (if there are no seasonal variations, fill in the first column) (1)

TYPES OF ANIMALS	NUMBER OF ANIMALS		
	Whole year	Winter	Summer
- Beef cattle - Dairy cattle - Calves - Swine - Sheep - Chickens - Horses - Other			

2. TYPE OF FEEDING (if there are no seasonal variations, fill in the first column) (2)

TYPE OF FEED	Whole year	Winter	Summer
. Raw fodder (hay - straw - pulp - grass - crop residues- other)			
. Compounds (soya - peanuts - pulp - cereals - rape - other)			
. Silage (maize - ray - grass - lucerne + clover - pulp - other)			
. Other (1)			

- Supplies of additives to the feed (2)
  . Cu ☐
  . antibiotics ☐
  . other (1)

3. HOUSING SYSTEM (1)

- Manure evacuation system (1) :

---

(1) specify
(2) mark the correct answer by crossing the corresponding square

ANNEX III : QUESTIONNAIRE FOR BIOMETHANATION ON LANDFILLS

1. LANDFILL (1)

- Dimensions - surface : ha
  - depth : m
- Age of the landfill : years
- Quantities of wastes : tons
- Mean composition of the wastes :

Nature	Percent

- Pretreatment of the wastes :

2. DESCRIPTION OF THE GAS COLLECTION SYSTEM (1)

- Type : horizontal gas pipelines, vertical gas pipelines, other...
- Material used for the pipes :
- Size of the pipes :
- Layout of the gas collection system :
  . distance between the pipes : m
  . surface area exploited : ha

3. GAS PRODUCTION (1)

For each area exploited, and for each year of exploitation, give :

- the surface area : ha
- the gas production : m^3 gas/ton wastes
  or m^3 gas/m^3 wastes

4. USES OF THE GAS (1)

---

(1) specify

ANNEX IV : OPERATIONAL PROBLEMS

List the <u>major and minor</u> problems encountered during the digestion and mark the correct answer on the list of problems below.

- Problems are considered to be major (M.P.) when they result in temporary shut-down of the plant or souring of the digester mixed liquor.
- Problems are considered to be minor (m.p.) when, although they do not result in major disruption of the running of the plant, they are costly in terms of component replacement and/or man-hours involved in maintenance and servicing, or when they are potentially hazardous.

1. <u>FEEDSTOCK COMPOSITION, STORAGE AND HANDLING</u> (2)	M.P.	m.p.
- Variation in feed due to animal diet changes		
- Feed variation due to changes in factory operation		
- Groundwater or rainwater entry during storage		
- Compositional changes during storage (acidification,...)		
- Settlement during storage		
- Scum formation		
- Loss in volatile organic content		
- Methane production during storage		
- Odour problems		
- Other (1) :		
2. <u>LOADING PROBLEMS</u> (2)		
- Blockage of pipes due to unwanted materials in the waste (grit, hair...)		
- Precipitation in pipelines as a result of high alkalinity in the water		
- Failure of pumping device (electric, mechanical, etc.)		
- Operational problems with heating of feed		
- Build-up of deposit on heat exchanger surface causing ineffective heat transfer		
- Blockage of heat exchanger		
- Other (1)		
3. <u>DIGESTER OPERATION</u> (2)		
- Scum formation		
- Inefficient mixing systems		
- Short-circuiting of feed flow in plug-flow systems		
- Settlement in the reactor		
- Clogging of upflow filters		
- Failure or inefficient heat transfer in heated digesters		
- Washout of active biomass		
- Problems with sludge granule stability or growth within the reactor		
- Difficulties with gas separation systems		
- Loss of active film in attached film reactors		
- Gas leakage		
- Problems with excess pressure release systems		
- Other (1)		

(1) specify
(2) mark the correct answer by crossing the corresponding line and column

	M.P.	m.p.
4. CHEMICAL/BIOCHEMICAL PROBLEMS (2)		
- Toxicity and inhibition		
. cations (ammonia, etc.)		
. metals		
. sulphide toxicity		
. presence of antibiotics		
- Inhibition due to the presence of unusual chemicals (*i.e.* chloroform)		
- Buffering problems		
- Other (1)		
5. GAS COLLECTION AND STORAGE (2)		
- Water blockage of gas lines		
- Freezing of H_2O in lines during winter operation		
- Leakage of storage systems		
- CH_4 diffusion through pipes or storage unit		
- Sticking of floating lid units		
- Problems with metering devices		
- H_2S problems		
- Compressor problems during gas pumping		
6. GAS UTILIZATION (2)		
- Gas stripping problems		
- Direct burning systems :		
. problems with design and construction material of burner		
. problems with flame-trap		
. problems with heat exchanger systems		
- Electricity generation systems		
. engine problems		
. difficulties with waste heat recovery systems		
. problems with off-peak use of electricity		
- Other (1)		
7. EFFLUENT HANDLING PROBLEMS (2)		
- Storage problems		
- Disposal problems		
. irrigation		
. aerobic final treatment		
. nitrogen removal		
. odour problems		
. pathogen carryover		
- Other (1)		

---

(1) specify
(2) mark the correct answer by crossing the corresponding line and column

8. GENERAL PROBLEMS (2)

Item	M.P.	m.p.
- Materials		
. corrosion of digester units and component parts		
. problems with insulation materials		
. gas storage bags		
. matrix decomposition in filters		
. electrical failures		
. other (1)		
- Monitoring systems		
. problems with temperature probes		
. pH probes		
. gas metering devices		
. other (1)		

---

(1) specify
(2) mark the correct answer by crossing the corresponding line and column

ANNEX V : TABLE OF SYMBOLS

TS : Total solids

VS : Volatile solids

COD : Chemical oxygen demand

TOC : Total organic carbon

BOD : Biochemical oxygen demand

N_{TK} : Total Kjeldahl nitrogen

N_{NH4^+} : Ammoniacal nitrogen

VFA : Volatile fatty acids

Pi : Inorganic phosphor

ML : Mixed liquor

d : Day

min : Minute

index

o : Influent

r : Removed

APPENDIX II.A. R & D PROGRAMMES IN THE MEMBER STATES OF THE EUROPEAN COMMUNITY AND SWITZERLAND

Research and development programmes on biomethanation are going on in every country of the European Community. These programmes were outlined in Part II, Chapter V. More detailed topics are listed below.

a) R & D Programme in Belgium

1° Applied research

Optimization of the anaerobic digestion of

- Agricultural wastes	
. cattle manure	: Unité de Génie Biologique, UCL(1)
. cattle manure with bedding	: Unité de Génie Biologique, UCL
. pig manure	: Lab. Microbiële Ecologie, RUG
. tropical wastes and residues	: Unité de Génie Biologique, UCL, Association Internationale de Développement Rural (AIDR).
. poultry manure	: Centre Agronomique de Recherches Appliquées du Hainaut (CARAH)
- Energy crops	
. freshwater and marine algae	: Unité de Génie Biologique, UCL
- Industrial wastes	
. bovine paunch manure	: Unité de Génie Biologique, UCL
. spent liquor from citric acid production	: Unité de Génie Biologique, UCL
. distillery and flax industry wastewaters	: Lab. Microbiële Ecologie, RUG, Studiebureau O. De Koninckx.
- Domestic refuse	: Cebedeau; Dep. Microbiologie, ULg; Lab. Microbiële Ecologie, RUG; Unité de Génie Biologie, UCL.
- Solid residues (composts, market refuse...)	: Lab. Microbiële Ecologie, RUG, Studiebureau O. De Koninckx, ACEC, (Ateliers de Construction Electrique de Charleroi), Unité de Génie Biologique, UCL.
2° Modelling	: Lab. Automatique et Analyse, UCL.
3° Economic analysis	: Faculté Notre Dame de la Paix de Namur

(1) See list of addresses, Appendix IV.

b) R & D Programme in Denmark

1° Technological development of biogas plants by the STUB project group. The latter is a joint research programme of nine research institutes and companies with the following aims :

- erection and testing of three full-scale demonstration biogas plants on existing farms,
- measurements and tests on existing biogas plants in Denmark,
- experiments with, and development of, process technology,
- experiments on, and assessments of, the fertilizer value of the digested sludge, the parasite kill-off and other possible environmental benefits,
- an information service on agricultural and technological topics for farmers and small industries as well as general information for the public.

2° Thermophilic and mesophilic anaerobic digestion of slaughter-house waste by the Aalborg University Centre :

- bench-scale continuous flow and batch methane digestion of contents from stomach and intestines (paunch manure) from slaughter-houses,
- at both mesophilic and thermophilic temperature regimes,
- with full-scale testing of laboratory results (partly financed by funds of the Commission of the European Communities).

3° Utilization of biogas in internal combustion engines by the Laboratory of Energetics of the Technical University of Denmark :

- dual fuel principle,
- spark ignition in a rebuilt diesel engine.

4° Investigations on the kinetics of microbial formation of methane by the Institute of Technical Biochemistry of the Technical University of Denmark :

- description of the conversion of organic material to methane,
- investigation of the possibility to control the microbial reaction (inter alia, influence of hydrogen partial pressure).

5° Biogas production from crop residues by the Royal Veterinary and Agricultural University : laboratory and economic investigations.

6° Status of the running conditions of existing biogas plants in Denmark by the Danish Biogas Technical Association : technical and economic evaluation of the possibilities of bringing the biogas plants to a reasonable operation level (accomplished in close cooperation with the owners of the biogas plants).

7° Development of motor/generator unit by the DEL Manufacturers of Electrical Equipment.

8° Development of biogas plants by Højbogard Biogas Plant Nørre Aby A/S :

- experiments with hardware;
- development of a biogas plant for commercial utilization.

9° Establishment of a cooperative biogas system (Revninge Biogas System) for the 100-family village of Revninge near Kerteminde based on agricultural waste (project partly financed by the Commission of the European Communities as a demonstration project.

10° Biogas from landfills by Crone & Koch, consulting engineers :

- investigation of the possibilities to utilise biogas from landfills in Denmark,
- outline of landfills, economic evaluation of plants and introduction to a pilot project.

11° Solid manure digestion in a continuous biogas plant by the Godrum project.

c) R & D Programme in the Federal Republic of Germany

1° Basic research

- Biochemical fundamentals of the formation of methane by the Fachgebiet Biotechnologie of Berlin, the Inst. f. Siedlungwasserwirtschaft (Univ. Hannover), and the Inst. f. Biotechnologie II of Jülich.

- Improvement of environmental conditions for the microorganisms in the different steps of decomposition by the 3 institutes cited above plus the Inst. f. Bodenbiologie and the Inst. f. Grünland- und Futterpflanzenforschung (FAL, Braunschweig).

- Influence of bactericides (antibiotics, detergents) on the process of decomposition, by the Botan. Inst. (Univ. Munchen),

- Multi-stage process, by Lehrstuhl B f. Thermodynamik (Univ. Munchen), the Inst. f. Grensflächen und Bioverfahrenstechnik of Stuttgart and the Inst. Grünland- und Futterpflanzenforschung (FAL, Braunschweig),

2° Technical research aimed at improving the performance of the methane digester

- Substrate specific material flow control (separating the components, enrichment of biomass, differentiated mean detention time), by the Inst. f. Technologie (FAL, Braunschweig),

- Rheology, by the Inst. f. Verfahrenstechnik (RWTH Aachen) and the Inst. f. Technologie (FAL, Braunschweig),

- Stirring and mixing, by the Inst. f. Agrartechnik/Landesanstalts f. Landw. Maschinen- u. Bauwesen (Univ. Hohenheim) and the Inst. f. Technologie (FAL, Braunschweig),

- Heat transfer, by the Inst. f. Technologie (FAL, Braunschweig),

- Control and regulation functions, by the Fachgebiet Biotechnologie of Berlin and Lehrstuhl B f. Thermodynamik (Univ. München).

3° Process research :

- Gas potential of various substrates by the Inst. f. Agrartechnik/Landesanstalt f. Landw. Maschinen- u. Bauwesen (Univ. Hohenheim) and the Inst. f. Technologie (FAL, Braunschweig),

- Influence of the running conditions on the digestion process, by the two laboratories cited above and the Inst. f. Siedlungwasserwirtschaft (Univ. Hannover), the Inst. f. Landw. Verfahrenstechnik (Univ. Kiel) and the Botan. Institut (Univ. München),

- Inactivation of pathogenic organisms during the biomethanation process by the Inst. f. Tiermedizin u. Tierhygiene (Univ. Hohenheim).

4° Research in economics :

- Evaluation of biogas plants using technical-economic criteria, by the Inst. f. Agrartechnik/Landesanstalt f. Landw. Maschinen u. Bauwesen (Univ. Hohenheim) and the Inst. f. Technologie (FAL, Braunschweig),

- Relative advantages of alternatives for end-uses of biogas, by the Inst. f. Landtechnik d. Landw. Fakultät/Bayr. Landesanstalt f. Landtechnik (Univ. München) and the Inst. f. Technologie (FAL, Braunschweig),

- Profitability of biogas plants within agricultural exploitations, by the two institutes cited above and the Inst. f. Betriebswirtschaft (FAL, Braunschweig).

5° Utilization of biogas

- Preparation of the biogas (cleaning, compression), by the Inst. f. Agrartechnik/Landesanstalt f. Landw. Maschinen u. Bauwesen (Univ. Hohenheim),

- Storage of gas and heat, by the Inst. f. Agrartechnik/Landesanstalt f. Landw. Maschinen u. Bauwesen (Univ. Hohenheim) and the Inst. f. Technologie (FAL, Braunschweig),

- Technology of gas engines, by the two institutes cited above,

- Technology of burning and compound energy systems.

6° Utilization of effluent

- Effect as fertilizer by Inst. f. Agrartechnik/Landesanstalt f. Landw. Maschinen - Bauwesen (Univ. Hohenheim), by Lehrst. f. Pflanzenernährung und Bodenkunde (FAL, Braunschweig),

- Odour reduction, by the Inst. f. Landw. Verfahrenstechnik (Univ. Kiel).

d) R & D Programme in France

This programme is coordinated by the AFME (Agence Française pour la Maîtrise de l'Energie)

1° Basic Research

- Isolation and characterization of hydrolytic fermentative, acidogenic and acetoclastic bacteria : INRA - Villeneuve d'Ascq

- Biochemistry and microbiology of methanogenesis	: University - Marseille, ORSTOM - Vert le Petit
- Genetics of bacteria	: Institut Pasteur - Paris

2° Applied research

- Biomass and agricultural wastes	
. algal biomass	: University Lille I INRA - Narbonne INRA - Villeneuve d'Ascq IRCHA - Montpellier
. physical limits of substrates for methanogenesis	: IRH - Nancy
- Animal wastes	
. cattle manure	: INRA & IRAT - Montpellier IRCHA - Vert le Petit INRA - Rennes INSA - Toulouse, with Elf Aquitaine ITCF - Boigneville
. pig wastes	: CEMAGREF - Antony EMC - Aspach ITP - Toulouse, with Air Liquide
. poultry wastes	: IRCHA - Vert le Petit
- Municipal wastes	
. household refuse	: Elf Aquitaine INRA - Villeneuve d'Ascq University of Languedoc, with Carene & Valorga
- Industrial wastes	
. potato processing	: IRCHA - Vert le Petit ORSTOM - Paris
. coffee pulps	: IFCC - Montpellier
. tannery wastes	: IRCHA - Vert le Petit, with CT Cuir
. canning factory wastewaters, wood derivates	: INRA - Villeneuve d'Ascq
. distillery	: INRA - Narbonne, with SGN
. milk serums	: ITG - Rioz
. sugar industries (wastewaters and pulps)	: IRIS - Villeneuve d'Ascq
. beet pulps	: University - Compiègne
. brewery wastes	: Bertin - Tarnos
. slaughter-house wastes	: INPL - Nancy University - Caen
. technology of reactors	: Degremont and Sté Lyonnaise des Eaux et de l'Eclairage Omnium de Traitement et de Valorisation, Saint-Gobain Techniques Nouvelles

- Economics
  . rural economics, system studies INRA - Grignon

e) R & D Programme in Greece

1° Applied research on the biomethanation of pig manure, poultry manure and olive oil wastewaters, by the Laboratory of Agricultural Structures of Athens.

2° Construction and monitoring of an industrial digester on vinassa from the alcohol production industry, by the Laboratory of Organic Chemistry and Technology of the Polytechnics of Athens.

f) R & D Programme in Ireland

1° Biomethanation of agricultural wastes and industrial effluents, by the Microbiology Department of the University College of Galway.

2° Anaerobic digestion of sewage sludges, by the Civil Engineering Department of the University College of Dublin.

3° Conventional digestion of pig slurry and full-scale demonstration of cattle waste, by An Foras Taluntais (Irish Agricultural Research Institute) and Bailieboro Co-operative, Ltd.

g) R & D Programme in Italy

National programmes on biomethanation do not exist in Italy. Nevertheless several universities and industries are interested in research on biomethanation. These research programmes are funded by CNR (National Council of Research) and by MPI (Ministry of Public Instruction). Some public institutions like CNR and ENEA (Nuclear Energy and Renewable Energy) conduct their research in their own laboratories. Some industries have cooperative programmes with public scientific organizations. These R & D programmes may be summarized as follows :

1° Basic research

. Istituto di Ingeneria Agraria/Istituto di Chimica Agraria/Istituto di Microbiologia Agraria e Tecnica dell'Università di Milano : "Microbial, biochemical and technical aspects of biomethanation", project granted by MPI

. Istituto di Chimica Applicata e di Metallurgia, Università di Cagliari : "Growth of acidogenic bacteria and modelling of contact anaerobic digesters", project funded by MPI.

2° Applied research

- Agricultural wastes

. ESAV/ENEA/Istituto Meccanica Agraria, Università di Padova : "Anaerobic digestion of cow manure", granted by ESAV.

. Assoreni : "Anaerobic digestion of cow manure", "Anaerobic digestion in fixed bed reactor".

. Istituto di Meccanica Agraria e Meccanizzazione della Università di Firenze : "Farm energy balance".

. ESAVE : "Present use and possible upgrading of agricultural wastes", granted by CNR

. CRPA : "Special programme on biogas", granted by Regione Emilia Romagna.

. Istituto di Ingegneria Agraria/Istituto di Chimica Agraria/Istituto di Microbiologia Agraria e Tecnica della Università di Milano : "Demonstrative programme on the anaerobic digestion of cow-house wastes", funded by Regione Lombarda.

. Istituto di Ingegneria Agraria, Università di Milano : "Construction and monitoring of plants for biogas production by cow manure", project granted by CNR .

. ENEA/Istituto Meccanica Agraria, Università di Padova : "Monitoring of anaerobic digesters", granted by ENEA.

- Industrial wastes

. CNR (Istituto di Ricerca sulle acqueà - Bari : "Anaerobic digestion of olive-oil wastewater".

. Orbat Distillery/Istituto di Microbiologia Agraria e Tecnica, Università di Bologna : "Thermophilic anaerobic digestion of distillery waste".

. Eneco/Agip Nucleare : "Anaerobic digestion of tannery wastes".

. Associazione Chimici del Cuoio/Agip Nucleare s.p.a. "Anaerobic digestion of tannery waste".

h) <u>R & D Programme in The Netherlands</u>

1° Basic research

- Microbiology : University of Nijmegen
  University of Groningen

2° Applied research

- Agricultural wastes : Agricultural Univ. of Wageningen
  . manures — IMAG
  . solid wastes — IBVL, Wageningen

- Industrial wastewaters (UASB process)	: Agricultural Univ. of Wageningen ESMIL CSM Delft Technical University
- Domestic refuse	: TNO, Appeldoorn

i) R & D Programme in the United Kingdom

1° Basic research

- Bacteriology	: Rowett Research Institute, Aberdeen Food Research Institute, Norwich
- Microbiology and ecology	: Polytechnics of Wales, Pontypridd

2° Applied research on biomethanation of :

- Agricultural wastes	
. Cattle manure	: National Inst. for Research in Dairying, Shinfield, Reading
. Pig and cattle manure	: Rowett Research Institute
. Green crops	: University College of Cardiff
- Industrial wastewaters	: Food Research Inst., Norwich Pollution Research Unit, UMIST Water Research Centre, Stevenage, Herts
- Domestic refuse	
. Biomethanation in digesters	: Warren Spring Laboratory
. Landfills	: Environmental Safety Group, AERE, Harwell, Oxon
3° Modelling	: Polytechnics of Wales
4° Research on biogas uses	: Leicester Polytechnics

j) R & D Programme in Switzerland

1° Improvement of the technology of biogas production and utilization with respect to net energy production, by the Swiss Federal Institute of Technology, ETH, Zürich (NEFF Biogas Project).

2° Concept of an energy self-sufficient cheese factory in combination with a pig fattening unit and biogas production, by the Milchtechnisches Labor. ETH, Zürich.

3° Development of a multi-stage fermentation process for the treatment of wastewater from the food processing industries, by the Federal Institute of Technology of Lausanne.

4° Integration of biogas plants in Swiss farms. Improvement of the fermentation process with respect to net energy production. Biogas utilization with diesel engines, by the Swiss Federal Research Station for Farm Management and Agricultural Engineering, FAT, Tänikon.

5° Quality of digested manure as a fertilizer, by the Swiss Federal Research Station for Agricultural Chemistry and Environmental Hygiene, FAC, Liebefeld.

6° Digestion of domestic refuse, by the University of Neuenburg.

7° Anaerobic digestion of night soil in developing countries, by the International Reference Center EAWAG.

APPENDIX II.B. SUPPLEMENTARY TABLES

The supplementary tables given in the following pages are referred to in Part II, Chapter I.

Table 2.1. Geographical distribution of full- and pilot-scale biogas plants in the European Community and in Switzerland according to the type of waste treated

Country	Type of waste				
	Agricultural	Energy crops	Domestic residues (landfills)	Industrial	Total
	Full + pilot scale	Pilot-scale	Full- + pilot-scale	Full- + pilot-scale	Full- + pilot-scale
Belgium	21 + 4			6 + 4	27 + 8
Denmark	22 + 1			3 + 3	25 + 4
F.R. Germany	75		10	12	97
France	62 + 12		2 + 3	10 + 5	74 + 20
Greece	3 + 1			1	4 + 1
Ireland	2 + 3			1 + 3	3 + 6
Italy	58 + 5	1	1	11 + 2	70 + 8
The Netherlands	21 + 1		3 + 8	22 + 1	46 + 10
United Kingdom	12 + 9		7 + 2	3 + 2	22 + 13
Switzerland	102 + 6				102 + 6
Total	378 + 42	1	23 + 13	69 + 20	470 + 76

Table 2.2. Number of biogas plants per type of liquid agricultural waste and per country

Type of waste	Belgium	Denmark	France	F.R. Germany	Greece	Ireland	Italy	The Netherlands	United Kingdom	Switzerland	TOTAL
Dairy cattle manure	1	10	2				13	10	2	34	72
Bovine cattle manure	8		6	34	1	1	5		1	8	64
Pig manure	9	8	21	18	1	4	40	4	8	31	144
Poultry manure	2		3	5							10
Rabbit manure	1										1
TOTAL	21	18	32	57	2	5	58	14	11	73	291

Table 2.3. Number of biogas plants per type of solid agricultural waste and per country

Type of waste	Belgium	France	F.R. Germany	The Netherlands	United Kingdom	TOTAL
Dairy cattle manure bedding		9				9
Bovine cattle manure bedding		15	7		1	23
Pig manure bedding		1	2			3
Poultry manure bedding	1	3				4
Rabbit manure bedding		1				1
Goat manure bedding		1				1
Sheep manure bedding		1				1
Horse manure bedding		1				1
Straw			1			1
Crop residues				1		1
TOTAL	1	32	10	1	1	45

Table 2.4. Number of biogas plants per type of mixed agricultural waste and per country

Type of mixed waste	Belgium	Denmark	France	F.R. Germany	Greece	Italy	The Netherlands	United Kingdom	Switzerland	TOTAL
Dairy cattle + pig manure	1	3	1				3	2	26	36
Bovine cattle + pig manure	1	2	1	4		1				9
Dairy cattle + poultry manure					1					1
Bovine cattle + poultry manure							1			1
Pig + poultry manure							2	1		3
Pig manure + cheesery wastewaters						1			1	2
Dairy cattle manure + crop residues								1		1
Pig manure + crop residues								4		4
Poultry manure + olive oil waste-water + other wastes					1					1
Dairy cattle + horse manure						1				1
Dairy cattle manure + wastewater of slaughter-house						1				1
Pig manure + dairy wastewater						1				1
Horse manure + garden residues									1	1
Pig manure + bovine cattle manure bedding			1	2						3
Bovine cattle manure bedding + poultry manure	1			1						2
Dairy cattle + pig + poultry manure							1		2	3
Manure + crop residues + waste-water + septic tank								1	1	2
Dairy cattle manure bedding + sheep manure bedding			1							1
Bovine cattle manure bedding + sheep manure bedding			2							2
Sheep manure bedding + goat manure bedding			1							1
Various manure bedding			1							1
TOTAL	3	5	8	7	2	5	7	9	32	77

Table 2.5. Number of biogas plants per type of industrial waste and per country

Type of waste	Belgium	Denmark	France	F.R. Germany	Greece	Ireland	Italy	The Netherlands	United Kingdom	TOTAL
Sugary wastewaters	5		1	7			2	8	1	24
Sugary wastes			1							1
Sugary + distillery wastewaters			1							1
Distillery wastewaters	1		3	1	1	1	10	1	1	19
Starch factory wastewaters			1			1		3	1	6
Pectin industry wastewaters		1	1	1						3
Confectionery wastewaters									1	1
Slaughter-house wastewaters		2						1		3
Slaughter-house wastes	2		2							4
Milk processing wastewaters						2				2
Potato-processing wastewaters		2	1					8		11
Brewery wastewaters	1		2							3
Flax industry wastewaters	1									1
Vegetable canning wastewaters			1							1
Olive oil wastewaters							1			1
Enzyme industry wastewaters			1							1
Citric acid ind. wastewaters									1	1
Leather ind. wastewaters				1						1
Wooden plates ind. wastewaters				1						1
Pickles canning wastewaters				1						1
Paper ind. wastewaters								1		1
Fish canning wastewaters								1		1
Onion processing wastewaters		1								1
TOTAL	10	6	15	12	1	4	13	23	5	89

Table 2.6. Evolution of the number of biogas plants per year

Year	Agricultural wastes	Energy crops	Domestic residues	Industrial wastes	TOTAL
1952	1				1
1956	1				1
1967	1				1
1969				1	1
1970				1	1
1973	1				1
1974	1				1
1975	1			2	3
1976	5			3	8
1977	3			3	6
1978	12			3	15
1979	29		1	5	35
1980	87		2	16	105
1981	107		7	20	134
1982	113	1	8	18	140
1983*	27		11	15	53
Not known	31		7	2	40

* Known data at beginning of 1983

Table 2.7. Distribution of agricultural biogas plants among the different classes of constructors, in percentage share of total

Class of constructor	Belgium	Denmark	France	F.R. Germany	Greece	Ireland	Italy	The Netherlands	United Kingdom	Switzerland
Owner	29	5	33	28			5		33	8
Company	65	23	42	63	33	100	82	85	58	89
Local tradesmen	6	5	6		33			5		
Owner + local tradesmen		18	4		33		12	10	8	
Owner + Company		50	10							3
Unknown			4	9						
	100	100	100	100	100	100	100	100	100	100

Table 2.8. Mean size of the agricultural biogas plants per class of constructor and per country, in m^3 total volume of methane digester

	Class of constructor				
Country	Company	Owner	Local tradesmen	Owner + Company	Owner + local tradesmen
Belgium	199	129	136	50	47
Denmark	280	56	150		
France	145	152	48	72	55
F.R. Germany	144	63	180		
Greece	300		200		180
Ireland	137				
Italy	846	189			
The Netherlands	263	331			
United Kingdom	366	123		34	
Switzerland	142	148		97	

APPENDIX III. BIOGAS PLANTS IN THE EUROPEAN COMMUNITY AND SWITZERLAND

The following listing gives a classification and a brief description of the 546 biogas plants located in the European Community and in Switzerland. The latter are classified as follows :

1° Per type of digested feedstock
   I. Agricultural wastes (liquid, solid, mixed),
   II. Energy crops,
   III. Domestic residues,
   IV. Industrial wastes.

A detailed breakdown of the listing is given on the following page.

2° Per country within each type of feedstock (the breakdown by country is indicated by horizontal lines)
- Belgium (B)
- Federal Republic of Germany (D)
- Denmark (DK)
- France (F)
- Greece (GR)
- Italy (I)
- Ireland (IRL)
- The Netherlands (NL)
- United Kingdom (UK)
- Switzerland (CH)

For each biogas plant, the following information is given whenever it is available :

1° The address of the biogas plant, <u>i.e.</u> its location.
2° The position of the plant on the maps presented in Part II, Chapter I.
3° The scale (full or pilot), and the total volume of the methane digester or the surface exploited for landfills.
4° The type of biomethanation system used with if possible, indications of the temperature maintained inside the methane digester.
5° The constructor of the biogas plant, who can be either a company for commercial constructors (see Appendix IV), or local tradesmen, or the owner in the case of a "do it yourself" construction.
6° The monitoring of the methane digester which, if it exists, can be carried out by a laboratory, an institute, a university or other (see Appendix IV).
7° The start-up, which indicates the date when the methane digester was filled for the first time. A second date can be given in the case of the break-down of a methane digester. There is only one exception for the industrial biogas plants of the Federal Republic of Germany for which the date of the start-up means the date of the first use of the biogas and is the only one available.

One remark should be added concerning the agricultural methane digesters of the Federal Republic of Germany. No distinction could be made between dairy cattle and bovine cattle manure so that the methane digesters working on these two types of manure are classified under bovine cattle manure.

<u>BREAKDOWN OF THE LISTING</u> <u>SERIAL Nos</u>

The serial numbers indicated for each sub-class of digester refer to the first column of the listing. To facilitate reference to a particular plant, the first and last serial numbers are printed at the top of each page.

I.	<u>Agricultural wastes</u>	001-420
I.1.	Dairy cattle manure	001-072
I.2.	Bovine cattle manure	073-136
I.3.	Pig manure	137-280
I.4.	Poultry manure	281-290
I.5.	Rabbit manure	291
I.6.	Dairy cattle manure bedding	292-300
I.7.	Bovine cattle manure bedding	301-323
I.8.	Pig manure bedding	324-326
I.9.	Poultry manure bedding	327-330
I.10	Rabbit manure bedding	331
I.11.	Goat manure bedding	332
I.12.	Sheep manure bedding	333
I.13.	Horse manure bedding	334
I.14.	Straw	335
I.15.	Crop residues	336
I.16.	Dairy cattle + pig manure	337-372
I.17.	Bovine cattle + pig manure	373-381
I.18.	Dairy cattle + poultry manure	382
I.19.	Bovine cattle + poultry manure	383
I.20.	Pig + poultry manure	384-386
I.21.	Pig manure + cheese dairy wastewater	387-388
I.22.	Dairy cattle manure + crop residues	389
I.23.	Pig manure + crop residues	390-393
I.24.	Poultry manure + olive oil wastewaters + other wastes	394
I.25.	Dairy cattle + horse manure	395
I.26.	Dairy cattle + wastewaters of slaughter-house	396
I.27.	Pig manure + dairy wastewaters (+ sludges)	397
I.28.	Horse manure + garden residues	398
I.29.	Dairy cattle + pig + poultry manure	399-401
I.30.	Manure + crop residues + wastewaters + septic tank	402-403
I.31.	Poultry manure bedding + bovine cattle manure bedding	404-405
I.32.	Pig manure bedding + bovine cattle manure bedding	406-408
I.33.	Dairy cattle manure bedding + sheep manure bedding	409
I.34.	Bovine cattle manure bedding + sheep manure bedding	410-411
I.35.	Sheep manure bedding + goat manure bedding	412
I.36.	Various manure bedding	413
I.37.	Type of agricultural waste unknown	414-420

I. AGRICULTURAL WASTES

	Address	Map	Scale (Size)	System	Constructor	Monitoring	Start-up
	I.1. DAIRY CATTLE MANURE						
001	Nouprez B-6688 ANCÔMONT (Lierneux)	E3	Full scale (50 m^3)	Completely mixed Int. loaded (35 °C)	Owner	UCL-GEBI Louvain- la-Neuve	10.1981
002	Bustrup Efterskole DK-7862 BRODAL	B3	Full scale (2 x 100 m^3)	Batch			1983
003	Ullerupvej, 5 DK-7323 GIVE	B4	Full scale (4 x 45 m^3)	Batch			
004	H. Andersen Forumvej, 121, Alslev DK-6732 GULDAGER	A5	Full scale (56 m^3)	Completely mixed Int. loaded (35-38 °C)	Owner	None	11.1980
005	Sdr. Vildstrup Bysvinget, DK-6100 HADERSLEV	B6	Full scale (2 x 150 m^3)	Continuous			1978-79 Closed
006	Assendrup Hovedgaard Assendrupvej, 10, Aversi DK-4690 HASLEV	E5	Full scale (2 x 220 m^3)	Plug-flow Int. loaded (33-35 °C)	Owner + local tradesmen	STUB Glostrup	08.1979
007	Holmagergärd Ørnevej, 2, Holmager DK-9240 NIBE	C2	Full scale (50 m^3)	Completely mixed Int. loaded (55-60 °C)	AGROMETAN Nibe	None	11.1981
008	Hoejbogaard Vibyvej, 27 DK-5580 Nr. AABY	C5	Full scale (4 x 40 m^3)	Batch			

	Address	Map	Scale (Size)	System	Constructor	Monitoring	Start-up
009	Roeding Hoejskole DK-6630 ROEDING	B5	Full scale (4 x 30 m^3)	Batch			1983
010	Enggaarden Stenderup DK-6681 HOLSTED	B5	Full scale (45 m^3)	Continuous			77-78 Closed
011	Nors Bangsmindesvej, 2, DK-7700 THISTED	B2	Full scale (4 x 45 m^3)	Batch			
012	Rolland F-73190 ST-JEOIRE-PRIEURE	G5	Full scale (100 m^3)	Completely mixed Int. loaded (35 °C)	METHANECO Montmelian	METHANECO Montmelian	End of 1979
013	Beaudoingt Les Alliberts F-38250 VILLARD-de-LANS	G6	Full scale (50 m^3)	Completely mixed Continuous	SIGOURE Roanne	None	
014	Cavis I-FELLIZANO (AL)	B2	Full scale (60 m^3)	Completely mixed Int. loaded	FM/ORAENERGIA Cacciola		04.1980
015	I-CEREGNANO (RO)	C2	Full scale (112 m^3)	Completely mixed Int. loaded (35 °C)	EMMEPI/GIZA Arbizzano	Universita di Padova	06.1981
016	Coop "L'AVANTI" I-CREMONA	B2	Full scale (240 m^3)	Completely mixed	ECOPLANTS S. Lazzaro		10.1982
017	Azienda Prella I-FOSSANO (CN)	A3	Full scale (70 m^3)	Completely mixed Int. loaded (37 °C)	SES Torino	Istituto Pro- fesionela Statale per l'Agricoltura di Cuneo	1979-1980

	Address	Map	Scale (Size)	System	Constructor	Monitoring	Start-up
018	Azienda Agricola F.lli Fochesato I-OLMO (VI)	C2	Full scale (150-200 m^3)	Completely mixed	EMMEPI Arbizzano		04.1982
019	Azienda F.lli Biolato I-SCARNAFIGI (CN)	A3	Full scale (60 m^3)	Completely mixed	EMMEPI Arbizzano		08.1982
020	Azienda Agricola Brandizzi I-TORRIMPIETRA (Roma)	D5	Full scale (600 m^3)	Completely mixed Int. loaded	SES Torino	SES Torino	11.1981
021	Stazione Sperimentale Agraria San Michele all Adige I-TRENTO	C1	Full scale (60 m^3)	Completely mixed	EMMEPI Arbizzano		08.1982
022	Azienda Agricola F.lli Invernizzi I-TRUCAZZANO (MI)	B2	Pilot scale (60 m^3)	Plug-flow	Owner	Istituto di Ingegneria Agraria Univ. Milano	
023	Azienda Agricola Menozzi I-LANDRIANO (PV)	C2	Pilot scale (2 x 26 m^3)	Batch	Owner	Istituto di Ingegneria Agraria Univ. Milano	
024	Azienda Agricola MENOZZI I-LANDRIANO (PV)	C2	Pilot scale (12 m^3)	Plug-flow	Owner	Istituto di Ingegneria Agraria Univ. Milano	
025	Azienda Agricola I-SCEVOLA (PV)	C2	Pilot scale (120 m^3)	Completely mixed	Owner	Istituto di Ingegneria Agraria Univ. Milano	

	Address	Map	Scale (Size)	System	Constructor	Monitoring	Start-up
026	Assoreni I-MONTEROTONDO (Roma)	D5	Pilot scale (3 m^3)			ASSORENI ENI Roma	
027	K.J. Doornbos S. Veldstraweg, 8 NL-ADUARD	D2	Full scale (148 m^3)	Completely mixed Int. loaded	PAQUES BV El-Balk	IMAG Wageningen	02.1982
028	Experimental dairy farm "De vijf roeden" Zesroedenweg, 1 NL-DUIVEN	D4	Full scale (120 m^3)	Completely mixed Int. loaded (30 °C)	WANROOY BV Heeswijk- Dinther	IMAG + IW-TNO + Agricultu- ral Univ. of Wageningen	11.1979
029	H. Griesen Apedijk, 5 NL-EIBERGEN	E4	Full scale (178 m^3)	Completely mixed Int. loaded	PAQUES BV El-Balk	IMAG Wageningen	01.1982
030	Heihoeve BV Heideweg, 3 NL-GIESSENBURG	C5	Full scale (600 m^3)	Completely mixed Int. loaded	OOSTWOUDER 'T Zand	IMAG Wageningen	01.1982
031	R.J. Dijkstra Soal, 8 NL-HEMELUM (Fr.)	C3	Full scale (200 m^3)	Completely mixed Int. loaded (32 °C)	PAQUES BV El-Balk	IMAG Wageningen	10.1980
032	R. Huitema Grevinnewei, 1 NL-JUTRIJP (Fr.)	C2	Full scale (350 m^3)	Completely mixed Int. loaded (34 $\pm$ 5 °C)	PAQUES BV El-Balk	IMAG Wageningen	09.1980
033	H. Willig Hogedijk, 8 NL-KATWOUDE	B4	Full scale (390 m^3)	Completely mixed Int. loaded	PAQUES BV El-Balk	IMAG Wageningen	01.1982

	Address	Map	Scale (Size)	System	Constructor	Monitoring	Start-up
034	Waiboerhoeve, Proefstation voor de Rundveehouderij Wisentweg, 55 NL-LELYSTAD	C3	Full scale (96 m^3)	Plug-flow Int. loaded (24-29 °C)	Fa. de BOER Leeuwarden	IMAG Wageningen	09.1980
035	H.J.M. Berendsen Revendijk, 3 NL-7137 H.J. LIEVELDE	D4	Full scale (220 m^3)	Plug-flow Int. loaded	OOSTWOUDER 'T Zand	IMAG Wageningen	01.1982
036	A. Van der Bosch Voorhaakdijk, 12 NL-WOERDENSE VERLAAT	B4	Full scale (148 m^3)	Completely mixed Int. loaded	PAQUES B.V. El-Balk	IMAG Wageningen	02.1982
037	Bore Place Farm Chiddingstone UK-EDENBRIDGE, Kent	G9	Full scale (300 m^3)	Plug-flow Int. loaded (34-40 °C)	HELIX M.P.S. Mortimer	ADAS	10.1979
038	NIRD, Nat. Inst. for Research in Dairying Shinfield UK - READING, Berks	F8	Full scale (2 x 130 m^3)	Completely mixed Int. loaded (20-40 °C)	SCS BIOTECH-NOLOGY Ltd. Stourport on Severn	NIRD-NIAE Poly. of Wales	04.1982
039	H. Hänni CH-8450 ADLIKON	D2	Full scale (250 m^3)	Fed-batch	LBA Brugg	None	02.1980
040	T. Hirschi Kirchbühl CH-3416 AFFOLTERN	C3	Full scale (47 m^3)	Completely mixed	ROTAVER AG Lützelfluh	None	
041	O. Müllhaupt CH-8962 BERGDIETIKON	D2	Full scale (55 m^3)	Completely mixed	WAHL Zollikon	None	1981

	Address	Map	Scale (Size)	System	Constructor	Monitoring	Start-up
042	A. Ueltschi CH-1304 DIZY	B3	Full scale (40 m^3)	Completely mixed	EPPARS Genève	None	1980
043	Inventa-Plarenga CH-7013 DOMAT-EMS	F3	Full scale (50 m^3)	Completely mixed	INVENTA AG Domat	None	03.1980
044	W. Röschmann Vogelsang CH-7013 DOMAT-EMS	F3	Full scale (50 m^3)	Completely mixed	INVENTA AG Domat	Owner	1978
045	H. Kaufmann CH-6182 ESCHOLZMATT	D3	Full scale (70 m^3)	Completely mixed	Owner	None	1980
046	L. Marido CH-2062 FENIN	B3	Full scale (55 m^3)	Completely mixed	LBA Brugg	None	05.1981
047	W. Früh CH-8376 FISCHINGEN	E2	Full scale (70 m^3)	Completely mixed	LEHMANN	None	10.1979
048	J. Wiederkehr CH-8911 JONEN	E3	Full scale (60 m^3)	Completely mixed	HUBER AG Lengnau	None	1981
049	M.E. Johner CH-3210 KERZERS	C3	Full scale (230 m^3)	Fed-batch (26 °C)	SCHWAB/PYTHON Kersers- Fribourg	None	1978
050	B. Rippstein Saalhöhe CH-4468 KIENBERG	C3	Full scale (60 m^3)	Completely mixed	HUBER AG Lengnau	None	1982
051	Pachtgut Bannholz CH-3098 KÖNIZ	C3	Full scale (200 m^3)	Fed-batch	CHAPALLAZ Vuiteboeuf	None	1982

	Address	Map	Scale (Size)	System	Constructor	Monitoring	Start-up
052	H.R. Wälti Holderacker CH-3438 LAUPERSWIL	C3	Full scale (55 m^3)	Completely mixed	Owner/ ROTAVER AG Lützelfluh	None	
053	Eidg. Forschungsanstalt für Agrikulturchemie und Umwelthygiene CH-3097 LIEBEFELD	C3	Full scale (60 m^3)	Completely mixed	LBA Brugg	Owner	01.1981
054	J.-C. Clerc CH-1349 MOIRY	B3	Full scale (100 m^3)	Completely mixed	PYTHON Fribourg	None	04.1980
055	M. Steiner CH-1171 MONTHEROD	A4	Full scale (75 m^3)	Completely mixed	HUBER AG Lengnau	Biogas Pro-jekt FAT-Tanikon	1976
056	Anliker CH-5037 MUHEN	D3	Full scale (72 m^3)	Completely mixed	ROTAVER AG Lützelfluh	None	03.1980
057	H.-U. Müller CH-8213 NEUNKIRCH	D1	Full scale (85 m^3)	Completely mixed	LBA Brugg	Biogas Pro-jekt FAT-Tanikon	02.1979
058	J. Konrad CH-8574 OBERHOFEN	E12	Full scale (80 m^3)	Completely mixed	HUBER AG Lengnau	Biogas Pro-jekt FAT Tanikon	10.1980
059	S. Chevalley CH-1599 PALEZIEUX	B3	Full scale (75 m^3)	Completely mixed	HUBER AG Lengnau	None	10.1980
060	F. Chevalley CH-1599 PALEZIEUX VILLAGE	B3	Full scale (100 m^3)	Completely mixed	HUBER AG Lengnau	None	09.1980

	Address	Map	Scale (Size)	System	Constructor	Monitoring	Start-up
061	J. Delafontaine CH-1604 PUIDOUX	B4	Full scale	Completely mixed	Owner	None	05.1980
062	Foyer d'éducation CH-2575 PRELES	B3	Full scale (260 m^3)	Completely mixed	LBA Brugg	None	01.1981
063	F. Haller CH-8911 RIFFERSWIL	D3	Full scale (50 m^3)	Completely mixed	LBA Brugg	Biogas Pro- jekt FAT- Tanikon	02.1979
064	Buholzer, Gebr. CH-6343 ROTKREUZ	D3	Full scale (140 m^3)	Completely mixed (30 °C)	LBA Brugg	None	1977
065	H. Blaser CH-6017 RUSWIL	E3	Full scale (69 m^3)	Completely mixed Int. loaded (35 °C)	ARNOLD AG Sempach- Station	None	02.1981
066	F. Rechsteiner CH-8374 SCHURTEN	E2	Full scale (50 m^3)	Completely mixed	LBA Brugg	None	09.1981
067	B. Varonne Branois CH-1950 SION	C2	Full scale (150 m^3)	Completely mixed	OLSOMMER Monthey	None	1982
068	R. Fuchs CH-5452 STARETSCHWIL	D3	Full scale (40 m^3)	Completely mixed	LBA Brugg	None	09.1980
069	W. Casparis CH-7430 THUSIS	F3	Full scale	Completely mixed	PRO ENERGIE	None	1980
070	H. Spycher CH-1111 VAUX S/MORGES	B4	Full scale (100 m^3)	Completely mixed	Owner	None	1977
071	F. Cornu CH-1099 VILLARS-MENDRAZ	B3	Full scale (85 m^3)	Completely mixed (16 °C)	BREMER	None	1978

	Address	Map	Scale (Size)	System	Constructor	Monitoring	Start-up
072	A. Rickenmann CH-8542 WIESENDANGEN	D2	Full scale (50 m^3)	Completely mixed	LBA Brugg	None	02.1981
			I.2. <u>BOVINE CATTLE MANURE</u>				
073	De Pierpont Rue de la Galoperie B-6464 AUBLAIN	C4	Full scale (32 m^3)	Plug-flow Int. loaded (30-35 °C)	Owner (ASBL 3 Vallées)	None	08.1981
074	J. Leroy Moulin Henrot B-6400 CHIMAY	C4	Full scale (100 m^3)	Completely mixed Int. loaded (33-35 °C)	Owner (F. Hardy)	UCL-GEBI Louvain- la-Neuve	1983
075	Marché de Ciney B-5300 CINEY		Pilot scale (20 m^3)	Completely mixed Int. loaded (35 °C)	Owner	AIDR/GEBI- UCL Louvain- la-Neuve	10.1979 to 07.1981
076	E. Flandre Rue de Froidchapelle B-6441 ERPION	C4	Full scale (8.4 m^3)	Accumulation Int. loaded (30 °C)	Owner (Fon- dation Rurale de Wallonie)	None	06.1980
077	Lab. de Génie Biologique (UCL), Place Croix du Sud, 1 bte 9 B-1348 LOUVAIN-LA-NEUVE		Pilot scale (750 l)	Plug-flow Int. loaded (35 °C)	Owner	GEBI-UCL Louvain- la-Neuve	01.1981 to 11.1981
078	Lab. de Génie Biologique (UCL), Place Croix du Sud, 1 bte 9 B-1348 LOUVAIN-LA-NEUVE		Pilot scale (2 m^3)	Accumulation Continuous (35 °C)	Local tradesmen	GEBI-UCL Louvain- la-Neuve	02.1980 to 06.1981
079	Albart B-6269 ROSELIES	D3	Full scale (255 m^3)	Completely mixed	ACEC Charleroi		02.1983

	Address	Map	Scale (Size)	System	Constructor	Monitoring	Start-up
080	L. Majoie Chaussée de Charleroi B-5921 THOREMBAIS-LES-BEGUINES	D3	Full scale (136 m^3)	Completely mixed Int. loaded (35 °C)	Local tradesmen	GEBI-UCL on behalf COMET Louvain-la-Neuve	02.1982
081	H. Schulze Mallinckrodtweg, 25 D-4730 AHLEN	B4	Full scale (50 m^3)	Completely mixed (35 °C)	AFAG Basel (CH)	None	12.1979
082	M. Haindl, Ederhof Arnstorfer str. 4 D-8353 ALTENMARKT	F8	Full scale (250 m^3)	Mixed (hydraulic) (33 °C)	BVT Bregenz (A)	Landtechnik Weihenstephan	12.1981
083	V. Reden-Lütcken Georgenau D-7263 BAD LIEBEN-ZELL-MOTTLINGEN	C8	Full scale (80 m^3)	Batch	Owner	Universität Hohenheim	1979
084	K. Abbt Sebastianstrasse, 1 D-8091 BATZENHOFEN (bei Augsburg)	D8	Full scale (42 m^3)	Completely mixed (35 °C)	Owner	None	05.1980
085	H. Schillingmann D-4576 BERGE DALVERS	B3	Full scale (68 m^3)	Plug-flow (15-30 °C)	Owner	FAL Braunschweig	03.1982
086	J. Kohnle D-7091 BERGHEIM (Tannhausen)	D7	Full scale (62 m^3)	Plug-flow + mixing (38 °C)	LIPP GmbH Tannhausen	None	11.1980
087	Bundesforschungsanstalt für Landwirtschaft (FAL) Bundesallee, 50 D-3300 BRAUNSCHWEIG	D4	Full scale (100 m^3)	Partially mixed (30-55 °C) Experimental	Owner	Inst. für Technologie Braunschweig	07.1980

	Address	Map	Scale (Size)	System	Constructor	Monitoring	Start-up
088	Hartwich u Bak Ingenieurburo S. Regenerative Verfahrenste Technik Am Schulberg 6a D-2114 DIERSTORF	C3	Full scale (100 m^3)	Completely mixed	REVAT Dierstorf	None	1983
089	R. Wurth Hohenlinderstrasse, 46 D-8016 FELDKIRCHEN	E8	Full scale		Owner	None	12.1982
090	A. Schierhölter Auerfehrden, 57 D-4519 GLANDORF	B4	Full scale	Plug-flow	BVT Bregenz (A)	None	1983
091	Staatl. Versuchgüterverwaltung Grub D-8011 GRUB	E8	Full scale (55 m^3)	Plug-flow (35 °C)	LIPP GmbH Tannhausen	Bayr. Landesanstalt für Tierzucht, Grub	06.1981
092	F. Keller Dissenhausen, 2 D-7959 GUTENZELL	D8	Full scale (450 m^3)		Owner	None	12.1982
093	H. Gries Loikum, 5 D-4236 HAMMINKELN	A4	Full scale (80 m^3)	Plug-flow (35 °C)	LIPP GmbH Tannhausen	Landwirtschaftskammer Rheinland Krefeld	04.1981
094	H. Behrens Berchersweg, 116, OT Tweelbäke D-2904 HATTEN	B3	Full scale (100 m^3)	Completely mixed (35 °C)	GEBKEN Hatten	Landwirtschaftskammer Weserems Oldenburg	12.1981

	Address	Map	Scale (Size)	System	Constructor	Monitoring	Start-up
095	K. Weitz Am Bügel, 7 D-6431 HOHENRODA- OBERBREITZBACH	C5	Full scale (76 m^3)	Plug-flow (36 °C)	Owner	Gesamthoch- schule Kassel + KTBL Darmstadt	12.1981
096	W. Hirrle Am Grenzwald, 1 D-7525 HOHENSTEIN BERNLOCH	C8	Full scale	Plug-flow	Owner	None	12.1982
097	G. Wagner D-8959 HOPFEN IM ALLGAU	CD9	Full scale (15 m^3)	Completely mixed (55 °C)	WLZ Saulgau	None	04.1979
098	W. Schraufstetter Schlosstr. 5 D-8045 ISMANING	E8	Full scale (2 x 600 m^3)	Plug-flow (35 °C)	Landwirt- schaftlicher Anlagenbau (W. Schrauf- stetter) Ismaning	None	08.1981
099	M. Ertl Panoramaweg, 1 D-KAY bei TITTMONING /Lkr. TRAUNSTEIN	E8	Full scale (45 m^3)	Plug-flow (35 °C)	Owner	None	01.1982
100	B. Barth Gutsverwaltung Erlenhof D-7527 KRAICHTAL/ UNTERÖWISHEIM	BC7	Full scale (100 m^3)	Completely mixed (35 °C)	BARTH, B. AGRARTECHNIK Kraichtal	Universität Hohenheim Stuttgart	1980
101	J. Stadler D-8170 OBERFISCHBACH (bei Bad Tölz)	E9	Full scale (300 m^3)	Plug-flow (37 °C)	Landwirt- schaftlicher Anlagenbau Ismaning	None	01.1981

	Address	Map	Scale (Size)	System	Constructor	Monitoring	Start-up
102	Betriebsgemeinschaft Ammertal, Unterdorfstr., 11 D-8121 ODERDING, GEMEINDE POLLING	D8	Full scale ($100\ m^3$)	Plug-flow (28 °C)	Landwirtschaftlicher Anlagenbau (W. Schraufstetter) Ismaning	None	06.1982
103	K.-H. LEINBERGER Neue Str. 1 D-6474 ORTENBERG 9, GELNHAAR	C6	Full scale ($80\ m^3$)	Completely mixed	BARTH, B. AGRARTECHNIK Kraichtal	Gesamthochschule Kassel + KTBL Darmstadt	06.1982
104	J. Brenner Kirchstr., 19 D-7095 RAINAU-DALKINGEN	D7	Full scale ($23\ m^3$)	Plug-flow (28 °C)	LIPP GmbH Tannhausen	None	01.1982
105	J. Gattinger D-8127 RETTENBERG	DE9	Full scale	Plug-flow	Landwirtschaftlicher Anlagenbau (W. Schraufstetter) Ismaning	None	12.1982
106	Wenning Altrhede, 3 D-4292 RHEDE	A4	Full scale ($490\ m^3$)	Completely mixed (34 °C)	HENZE-HARVESTORE GmbH, Hamm	None	02.1982
107	M.P. Hildebrandt D-2852 RINGSTEDT bei BREMERHAVEN	C2	Full scale ($80\ m^3$)	Plug-flow	Owner	Fachhochschule Bremerhaven	04.1981
108	J. Sedlmair D-8061 RUDELZHOFEN POST RÖHRMOOS	E8	Full scale ($50\ m^3$)	Plug-flow (40 °C)	Owner	Landtechnik Weihenstephan	04.1981

	Address	Map	Scale (Size)	System	Constructor	Monitoring	Start-up
109	A. Keller Krummengraben, 56 D-8121 ROTTENBUCH	D8	Full scale (100 m^3)	Completely mixed (32 °C)	BVT Bregenz (A)	None	12.1982
110	Carl-Duisberg Hof Gut Heiming D-8201 SCHECHEN	E89	Full scale	Completely mixed	LIPP GmbH Tannhausen	None	1980
111	F. Böse Päpsen, 1 D-2839 SIEDENBURG	C3	Full scale (20 m^3)	Completely mixed (35 °C)	Owner	None	03.1982
112	K. Weidenbacher Kreuthof D-7091 STÖDTLEN (bei TANNHAUSEN)	D7	Full scale (40 m^3)	Partially mixed (32 °C)	LIPP GmbH Tannhausen	None	01.1980
113	Kornbrennerei j. Grobrügge D-2723 WESTERVESEDE 4	C3	Full scale (300 m^3)	Plug-flow (32 °C)	BVT Bregenz (A)	None	03.1983
114	X. Königsbauer Parchetstr., 58 D-8120 WEILHEIM	D9	Full scale (40 m^3)	Completely mixed	Owner	Landtechnik Weihenstephan	12.1982
115	Lycée agricole F-COURCELLES-CHAUSSY	G2	Full scale	Completely mixed	Société BIOFLOR Nancy		1981
116	Jallifier La Chenevraie F-38250 LANS-en-VERCORS	G6	Full scale (44 m^3)		Sté CITREM		1982

	Address	Map	Scale (Size)	System	Constructor	Monitoring	Start-up
117	INA PG Domaine expérimental de l'INA F-78850 THIVERVAL GRIGNON	E3	Full scale 3 x 50 m^3	Batch	Owner	INA Thiverval-Grignon	1982-1983
118	INA PG Domaine expérimental de l'INA F-78850 THIVERVAL GRIGNON	E3	Full scale (35 m^3)	Plug-flow	Owner	INA Thiverval-Grignon	1982-1983
119	INA PG Domaine expérimental de l'INA F-78850 THIVERVAL GRIGNON	E3	Full scale (35 m^3)	Continuous	ZODIAC St-Cloud	INA Thiverval-Grignon	1982-1983
120	Société Civile Agricole du Domaine Ollwiller WUENHEIM 68500 F-68500 GUEBWILLER	H3	Full scale (120 m^3)	Completely mixed Continuous	MARUMO Guebwiller		1982
121	J. Sofianidis Alonaki GR-KOZANI	C2	Full scale (200 m^3)	Partially mixed Int. loaded	Local tradesmen		1983
122	Istituto Agrario Bonafous I-CHIERI	A2	Full scale (300 m^3)	Completely mixed Int. loaded	SES Torino	EUROACTIZYME	03.1981
123	Persico I-CREMONA	B2	Full scale (1000 m^3)	Completely mixed Int. loaded	SES Torino	SES Torino	1982
124	Azienda AG. S. Giorgio Spa I-POMBIA (NO)	A2	Full scale (400 + 200 m^3) Int. loaded	Completely mixed in series (35 °C)	ECOPLANTS Lazzaro	None	01.1982

	Address	Map	Scale (Size)	System	Constructor	Monitoring	Start-up
125	Azienda l'Ortolia I-REDEMELLO SOTTO (BS)	C2	Full scale (260 m^3)	Completely mixed Int. loaded (35 °C)	ECOPLANTS S. Lazzaro	None	01.1982
126	Vicchio I-MUGELLO (FI)	C4	Full scale (250 m^3)	Plug-flow	Owner	University of Firenze	07.1981
127	Johnston Castle IRL-Co. WEXFORD	C8	Full scale (240 m^3)	Completely mixed Int. loaded (35 °C)	PAQUES (NL) El-Balk	An Foras Taluntais Co-Wexford	03.1982
128	McFall, The Heaning Farm, Dunsop Bridge UK-NEWTON IN BOWLAND (Lancs.)	E5	Full scale (13 m^3)	Plug-flow	Owner	None	06.1982
129	H. Rüfenacht CH-1681 CHESALLES	B3	Full scale (47 m^3)	Completely mixed	ROTHEN Moudon	None	10.1981
130	P. Jaeggi CH-1249 COLLEX-BOSSY	A4	Full scale (36 m^3)	Plug-flow	PFLUG Cologny	None	1982
131	Landwirtschaftliche Schule Strickhof CH-8307 LINDAU		Pilot scale (6.85 m^3)	Upflow without carrier	Biogas Projekt ETH - Zentrum Zürich	Biogas Projekt ETH - Zentrum Zürich	1981
132	Landwirtschaftliche Schule Strickhof CH-8307 LINDAU		Pilot scale (8.16 m^3)	Partially mixed	Biogas Projekt ETH - Zentrum Zürich	Biogas Projekt ETH - Zentrum Zürich	1981

	Address	Map	Scale (Size)	System	Constructor	Monitoring	Start-up
133	Landwirtschaftliche Schule Strickhof CH-8307 LINDAU		Pilot scale (20 m^3)	Partially mixed (manure displaced through a coarse sieve)	Biogas Projekt ETH - Zentrum Zürich	Biogas Projekt ETH - Zentrum Zürich	1982
134	E. Kuratli CH-3053 MÜNCHENBUCHSEE	C3	Full scale (170 m^3)	Completely mixed	LBA Brugg	None	07.1981
135	M.P. Fischer CH-8439 RÜMIKON	D2	Full scale (314 m^3)	Fed-batch (12-20 °C)	LBA Brugg	Biogas Projekt FAT-Tanikon	11.1980
136	K. Meier CH-5305 UNTERENDINGEN	D2	Full scale (66 m^3)	Partially mixed (33 °C)	HUBER AG Lengnau	Biogas Projekt FAT-Tänikon	1978
				I.3. PIG MANURE			
137	Cordenier Lembekestraat, 8 B-9968 ASSENEDE BASSEVELDE	B1	Full scale (2 x 210 m^3)	Completely mixed Int. loaded (30-33 °C)	Owner	RUG-COVAL Gent	02.1982
138	Vanwilderode Monnikhofstraat, 3 B-1761 BORCHTLOMBEEK	C2	Full scale (2 x 50 m^3)	Plug-flow	BIOWATT Merchtem	RUG Gent	1983
139	Pilet, ferme de Chenemont B-6330 CORROY-LE-CHATEAU	D3	Full scale (270 m^3)	Plug-flow Int. loaded	AIDR Bruxelles	AIDR-GEBI Louvain-la-Neuve	1983
140	Van Haverbeke Schendekouterstraat, 19 B-8660 GELUWE	B2	Full scale (80 m^3)	Completely mixed Int. loaded (35 °C)	IVPO Mariakerke	IVPO-RUG Gent	02.1982

	Address	Map	Scale (Size)	System	Constructor	Monitoring	Start-up
141	Vaesen Overmaai, 18 B-9691 KERKHOVE	B2	Full scale (40 m^3)		EKW PVBA Eksel		1983
142	Commune de Sprimont B-4060 SPRIMONT	E3	Full scale ($\pm$ 1000 m^3)	Completely mixed		Institut de pathologie Univ. de Liège, Sart-Tilman	1983
143	Devos Pijpegale, 13 B-8981 LO RENINGE	A2	Full scale (300 m^3)	Completely mixed	IVPO Mariakerke	IVPO + RUG Gent	1983
144	Boerderij B-RIJKEVORSEL		Pilot scale (12 m^3)	Accumulation Int. loaded (33 °C)	Owner	RUG-COVAL Gent	09.1981
145	Decapmaker M. Sint Elooiweg, 47 B-8903 VOORMEZELE	A2	Full scale (87 m^3)	Completely mixed Int. loaded (30 °C)	Owner	RUG-COVAL Gent	03.1981
146	O. Greifenstein Breitfeldweg, 9 D-8351 AHOLMING	F8	Full scale (75 m^3)	Partially mixed (35 °C)	BVT Bregenz (A)	None	1982
147	V. Bornebusch Gut Wahrberg D-8801 AURACH	D7	Full scale (63 m^3)	Completely mixed (55 °C)	GFIT Roth	Universität Hohenheim Stuttgart	12.1982
148	K. Teeman Hausstette, 71 D-2849 BAKUM 6	B3	Full scale (90 m^3)	Plug-flow (33 °C)	LIPP GmbH Tannhausen	Landwirt- schaftskammer Weser-Ems Oldenburg	01.1982

	Address	Map	Scale (Size)	System	Constructor	Monitoring	Start-up
149	Joba-Biogas Postfach D-2830 BASSUM	C3	Full scale	Completely mixed	JOBA BIOGAS Bassum	None	
150	H. Fischer Glashütter Str, 26 D-3205 BOCKENEM 3	D4	Full scale (280 m^3)	Completely mixed (10-15 °C)	UTEC Göhrde	FAL Braunschweig	07.1981
151	Petersen Desmerciereskoog D-2257 BREDSTEDT	C1	Full scale (95 m^3)	Completely mixed (35 °C)	ÖKOTHERM Hagen	Institut für Landwirt- schaftliche Verfahrens- technik, Kiel	1979
152	G. Oltmann Aschenstedter Str., 17 D-2879 DOTLINGEN	B3	Full scale (200 m^3)	Completely mixed	GEBKEN	None	1983
153	M. Wallner; B. Fink Schulstrasse, 8 D-8031 GILCHING	D8	Full scale (90 m^3)	Completely mixed (33 °C)	MÜLLER Rosenheim	None	03.1980
154	Altes amt D-3355 KALEFELD	C4	Full scale (350 m^3)	Plug-flow (32 °C)	BVT Bregenz (A)	FAL Braunschweig	1983
155	P. Herlyn Kloster Sielmönken D-2974 KRUMMHORN 1	B23	Full scale (30 m^3)	Completely mixed (52 °C)	CHEMOTECHNICK Weiler	Landwirt- schaftskammer Weser-Ems Odenburg	12.1982
156	F. Eugen Eichensteiner Str. 21 D-8671 ISSIGAU (Lkr. Hof)	E6	Full scale	Plug-flow	BVT Bregenz (A)	Landwirt- schaftsamt Münchberg	1983

	Address	Map	Scale (Size)	System	Constructor	Monitoring	Start-up
157	O. Fuchs Schöllkrippener Str. 30 D-8752 KROMBACH	C6	Full scale (170 m^3)	Completely mixed (35 °C)	Owner	KTBL Darmstadt	12.1982
158	F. Hans Haupstr. 8 D-4149 LOUISENDORF	A4	Full scale (120 m^3)	Plug-flow (32 °C)	BVT Bregenz (A)	Landwirt- schaftskammer Rheinland Krefeld	10.1981
159	A. Schneider Hof am Storchennest D-8011 MICHELFELD	D7	Full scale (60 m^3)	Completely mixed (32 °C)	Gesellschaft für Alterna- tive Energieer- zeugung mbH Michelfeld	Landtechnik Weihenstephan	11.1980
160	H. Baumgartner Salzstrasse, 40 D-8000 MÜNCHEN 81 DORNACH	E8	Full scale (80 m^3)	Completely mixed (25 °C)	GEIGER München	None	01.1980
161	J. Sieh D-2371 RADE bei RENDSBURG	C2	Full scale	Plug-flow	Owner	Institut für Landwirt- schaftliche Verfahrens- technik, Kiel	1983
162	P. Eberle Stöckhof, 1 D-7253 RENNINGEN	C8	Full scale (78 m^3)	Plug-flow (35 °C)	LIPP GmbH Tannhausen	Universität Hohenheim Stuttgart	
163	A. Reinhardt Im Galgenberg D-8358 VILSHOFEN	F8	Full scale (60 m^3)	Plug-flow (35 °C)	LIPP GmbH Tannhausen	None	06.1981

	Address	Map	Scale (Size)	System	Constructor	Monitoring	Start-up
164	A. Flensburg, Damgärden Hjelselvej, 10, Hygum DK-7300 JELLING	B5	Full scale (3 x 60 m^3)	Batch (35 °C)	Owner + local tradesmen	None	08.1979
165	Overbyvej, 7 Skarild DK-6933 KIBAEK	B4	Full scale				Closed
166	DK-LEJRE	E5	Pilot scale (20 m^3)				
167	Sjoelundgaard Sjoelund Landevej, 24 DK-6093 SJOELUND	C6	Full scale (6 x 66.5 m^3)	Completely mixed Int. loaded (35 °C)	ANDERSEN & FOGLDD Risskov	STUB Glostrup	10.1979
168	Skadborg, Gadebjerg- gaard, Herredsvejen, 12 DK-8832 SKALS	B3	Full scale (410 m^3)	Completely mixed Int. loaded (35 °C)	JYSK BIOGAS Farsø	STUB Glostrup	09.1979
169	Bukkerupgaard Bukkerupvej, 64 DK-4340 TOELLOESE	E5	Full scale (300 m^3)				Closed
170	Hjelmerup Lindebjergsvej, 14 DK-5690 TOMMERUP	C6	Full scale (100 m^3)				
171	Godrum Trundballevej, 34 DK-8260 VIBY	C4	Full scale (50 m^3)				1983

	Address	Map	Scale (Size)	System	Constructor	Monitoring	Start-up
172	Mauget GAEC du printemps Le Grand Amgibou F-49600 BEAUPREAU	C4	Full scale (200 m^3)	Completely mixed	Sté WOLF		1982-1983
173	Institut Supérieur Agricole de Beauvais, B.P. 313 F-60026 BEAUVAIS	E2	Pilot scale (15 m^3)		Owner		1982
174	F-80 BEAUVAL	E2	Full scale (33 m^3)	Completely mixed Int. loaded (35 °C)	SSCM St-Etienne	SARL/CAMBIO	09.1981
175	Ferme expérimentale SANDERS, St-Symphorien F-72480 BERNAY-EN-CHAMPAGNE	D3	Full scale (250 m^3)	Completely mixed Int. loaded	PEC/ENGINEERING, EMC GROUP Paris	SCPA ENSAIA Nancy	07.1981
176	Mouthon, Les Marceaux St-Etienne Le Molard F-42130 BOEN	F5	Pilot scale (20 m^3)	Completely mixed Int. loaded (35 °C)	SSCM St-Etienne	SSCM St-Etienne	06.1980
177	GAEC Puy-Japin Viersat F-23170 CHAMBON-SUR VOUEZE	D5	Full scale		Sté SCTR		1982-1983
178	Cheva, Menglon F-26410 CHATILLON-EN-DIOIS	G6	Full scale (44 m^3)		Sté NRJ		1982-1983
179	Etablissements Ferard F-35210 CHATILLON-EN-VENDELAIS	C3	Full scale (1200 m^3)	Completely mixed Int. loaded (37-38 °C)	PEC ENGINEERING + Sté EVERS Paris	CEMAGREF Rennes	02.1980

	Address	Map	Scale (Size)	System	Constructor	Monitoring	Start-up
180	Pinel, La Baffrière F-35520 MELESSE (RENNES)	C3	Full scale (30 m^3)	Completely mixed Int. loaded (35 °C)	SSMA-SSCM St-Etienne	Chambre d'Agriculture + Féd. départementale des CUMA	10.1981
181	Hanne, La Rosière F-59170 MERIGNIES	E1	Full scale (4 x 25 m^3)	(55 °C)			1982
182	M. François Anolieu-Mepieu F-38390 MONTALIEU	G5	Full scale (3 x 38 m^3)	Completely mixed Int. loaded (25-30 °C)	Owner + METHANECO Montmelian	None	04.1980
183	Ledoux, Adane 5, Chemin de la petite verse F-60400 NOYON	E2	Full scale (40 m^3)	Completely mixed	Sté ARPEL		End of 1980
184	Daumeon Villegaudin F-22190 PLERIN	B3	Full scale (40 m^3)	Continuous			
185	Elevage Franquet père & fils, 4, rue de Paris F-95500 ROISSY-EN-FRANCE	E2	Full scale (200 m^3)	Completely mixed Int. loaded (40 °C)	Owner	INRA CEMAGREF ANRED	02.1982
186	Chaffel St-Sauveur-en Diois F-26340 SAILLANS	G6	Full scale (60 m^3)	Continuous	METHANECO Montmelian		1982
187	M. Boisset, Quartier des Sabilières-Clerieux F-26260 ST DONAT SUR L'HERBASSE	F6	Full scale (36 m^3)	Completely mixed Int. loaded (22-37 °C)	Owner + METHANECO Montmelian	None	10.1980

	Address	Map	Scale (Size)	System	Constructor	Monitoring	Start-up
188	Station expérimentale EDE du Finistère Trevarez F-29163 ST-GOAZEC	A3	Full scale ($30\ m^3$)	Completely mixed Int. loaded (35 °C)	Sté SSMA		1981
189	Chabagno Elevage d'Iratzia, Lasse F-64220 ST-JEAN-PIED-DE-PORC	C7	Full scale ($250\ m^3$)	Completely mixed Continuous	SAPEC ELF-AQUITAINE Pau		End of 1982
190	Socadexa, ferme de multiplication porcine F-68 SCHOENEN-STEINBACH	H3	Full scale ($232\ m^3$)	Completely mixed Int. loaded (36 ± 1 °C)	PEC ENGINEERING EMC GROUP Paris	SCPA ENSAIA Nancy	07.1981
191	Station expérimentale de l'Institut Technique du porc F-12200 VILLEFRANCHE-de-ROUERGUE	E6	Pilot scale ($3\ m^3$)	Anaerobic filter Continuous (35 °C)	AIR LIQUIDE Champigny sur Marne	ITP Toulouse	09.1980
192	Station expérimentale de l'Institut technique du porc F-12200 VILLEFRANCHE-de-ROUERGUE	E6	Pilot scale ($5 + 23\ m^3$)	Two-step Continuous (35 °C)	AIR LIQUIDE Champigny sur Marne	ITP Toulouse	09.1980
193	G. Lygizos Agios Thomas GR-VIOTIA	D4	Full scale ($160\ m^3$)	Plug-flow Int. loaded	Local tradesmen	Lab. of Agric. Structures, Agric. College of Athens	1983

	Address	Map	Scale (Size)	System	Constructor	Monitoring	Start-up
194	Azienda Agricola Comprozoo I-ASCOLI PICENO	D4	Full scale ($400\ m^3$)	Batch	EUROCOM S. Lazzaro		06.1982
195	SASSA Spa I-ASSEMINI (Cagliari)	B6	Full scale ($300\ m^3$)	Completely mixed Int. loaded	RPA Fontana	RPA Fontana	01.1982
196	Fratelli Testa I-BAGNACAVALLO (RA)	C3	Full scale ($260\ m^3$)	Completely mixed (35 °C)	AGIP-GIZA Spa, Reggio Emilia	CRPA-CNEN Reggio Emilia	02.1982
197	Societa Agricola Umbra Spa, Azienda di Brufa di Torgiano I-BASTIA UMBRA (PG)	D4	Full scale ($500\ m^3$)	Completely mixed Continuous	RPA Fontana	RPA Fontana	08.1981
198	Azienda di Bettona Societa Agricola Umbra Spa I-BASTIA UMBRA (PG)	D4	Full scale ($2000\ m^3$)	Completely mixed Continuous	RPA Fontana	RPA Fontana	09.1980
199	Pig-Italia Spa I-BASTIA UMBRA (PG)	D4	Full scale ($500\ m^3$)	Completely mixed Continuous	RPA Fontana	RPA Fontana	09.1981
200	Codep Bastia I-BASTIA UMBRA (PG)	D4	Full scale ($2000 + 500\ m^3$)	Completely mixed Continuous (series)	SNAMPROGETTI PESARO Fano	CRPA CNEN Reggio Emilia	1982
201	Codep-Bettona I-BETTONA (PG)	D4	Full scale ($7000 + 2500\ m^3$)	Completely mixed Continuous (series)	RPA Fontana	RPA Fontana	12.1982
202	Azienda Aras I-CADELBOSCO DI SOTTO (RE)	C3	Full scale ($320\ m^3$)	Completely mixed Continuous (35 °C)	AGIPGIZA Spa Reggio Emilia	ASSORENI Roma	1979

	Address	Map	Scale (Size)	System	Constructor	Monitoring	Start-up
203	Azienda Stella I-CAMERATA PICENA (AN)	D3	Full scale (400 + 400 m^3)	Completely mixed Two-step (33-37 °C)	RODON Rimini	Univ. di Bologna Padova Ancona	04.1982
204	Azienda Citterio I-CAROBBIO D.A. (BG)	B2	Full scale (900 + 450 m^3)	Two-step Continuous (35 °C)	ECOPLANTS S. Lazzaro	ECOPLANTS S. Lazzaro	12.1980
205	Azienda Agricola S. Anna I-CASTELLAZZO (NO)	A2	Full scale (1800 m^3)	Completely mixed	EMMEPI Arbizzano		1982
206	Latterie Sociali Coop I-CHIARI	C2	Full scale (1650 m^3)	Batch	EUROCOM S. Lazzaro		05.1981
207	Agricola Todini Spa I-COLLEVALENZA DI TODI (PG)	D4	Full scale (2 x 1000 m^3)	Completely mixed Continuous	RPA Fontana	RPA Fontana	09.1979
208	Consorzio Agrario Pro- vinciale Ferrara Azienda Agraria "Bellario" I-COMACCHIO (FE)	C2	Full scale (300 m^3)	Completely mixed Int. loaded	RPA Fontana	RPA Fontana	07.1980
209	Caseificio Crescentino I-CRESCENTINO (VC)	C2	Full scale (100 m^3)		ECOTECNICA		1980
210	Azienda Agricola Rostagno Teresio I-FOSSANO DI CUNEO (CN)	A3	Full scale (300 m^3)	Completely mixed Int. loaded	RPA Fontana	RPA Fontana	01.1981
211	Azienda Autostrada del Sole I-GAVASSA (RE)	C3	Full scale (1200 m^3)	Completely mixed Continuous	AGIPGIZA Reggio Emilia	AGIPGIZA Reggio Emilia	07.1981

	Address	Map	Scale (Size)	System	Constructor	Monitoring	Start-up
212	Azienda Agricola Pianoni I-IESOLO LIDO	C2	Full scale (530 m^3)	Completely mixed	EMMEPI Arbizzano		1980
213	Azienda Agricola "S. Bernardo" I-LESEGNO (CN)	A3	Full scale (1000 + 400 m^3)	Completely mixed Continuous (in series)	RPA Fontana	RPA Fontana	06.1982
214	Ditta Zoomer I-LUCERA (FG)	F5	Full scale (600 m^3)	Completely mixed	FM/ORAENERGIA Spa		09.1982
215	Azienda Agraria "Sturla" I-MANERBIO (BS)		Full scale (800 m^3)	Batch	EUROCOM S. Lazzaro		06.1981
216	Azienda S. Francesco I-MELDOLA (FO)	C3	Full scale (670 m^3)	Anaerobic contact Continuous (35 °C)	AGIPGIZA Reggio Emilia	CRPA CNEN Reggio Emilia	02.1981
217	Azienda Amadori I-MERCATO SARACENO (FO)	C3	Full scale (658 m^3)	Completely mixed Continuous (35 °C)	TECNOIMPIANTI Montiano	CRPA CNEN Reggio Emilia	02.1982
218	Azienda Agricola La Posta I-MERCORE (PC)	B3	Full scale (1300 m^3)	Completely mixed Continuous	BIOSYSTEM S.R.L. Milano	MONTEDISON	01.1981
219	Invernizzi Spa Stabilimenti di Liscate I-MILANO	B2	Full scale (2500 m^3)	Completely mixed Int. loading (35 °C $\pm$ 1 °C)	AGIPGIZA Reggio Emilia		1983
220	Azienda Vittori I-MINERBIO (BO)	C3	Full scale (600 m^3)	Completely mixed Continuous	AGIPGIZA Reggio Emilia	AGIPGIZA Reggio Emilia	11.1981
221	Caseificio case Bonazzi I-MONTE SAN PIETRO (BO)	C3	Full scale ($\pm$ 12 x 17 m^3)	Completely mixed	Owner (Camera Commercio BO)	Centro Ricer-che FIAT Orbassano	09.1981

	Address	Map	Scale (Size)	System	Constructor	Monitoring	Start-up
222	Latteria Vo Grande I-PEGOGNAGA (MN)	B2	Full scale (50 m^3)	Completely mixed	EURODEPURATORI Mantova	None	10.1980 to 01.1982
223	Azienda Zavatta I-POGGIO BERNI (FO)	C3	Full scale (900 m^3)	Completely mixed Int. loaded (35 °C)	EMMEPI Arbizzano	CRPA-CNEN Reggio Emilia	02.1981
224	Azienda ICSA I-SOLIERA (RE)	C3	Full scale (1200 m^3)	Completely mixed Continuous (35 °C)	CCPL EBiA Milano	CRPA-CNEN Reggio Emilia	1982
225	Corpo Forestale Stato I-RADICONDOLI (SI)	C4	Full scale (300 m^3)	Completely mixed	FM/ORAENERGIA Spa		07.1982
226	Bonacini I-SCANDIANO (RE)	C3	Full scale (60 m^3)	Completely mixed	FM-ORAENERGIA Spa		02.1980
227	Azienda Rusticalli I-S. PIETRO IN BAGNO (FO)	C3	Full scale (450 m^3)	Completely mixed Int. loaded (35 °C)	CLOVER Forli		12.1980
228	Azienda Gasparini I-S. PROSPERO di MODENA (MO)	C3	Full scale (488 m^3)	Completely mixed Int. loaded (35 °C)	SNAMPROGETTI PESARO Fano	CRPA-CNEN Reggio Emilia	1982
229	Center Pig I-TODI (PG)	D4	Full scale (250 m^3)		BSF Roma		1982
230	La Zootecnica Umbra I-TODI (PG)	D4	Full scale (120 m^3)		BSF Roma		1980
231	Azienda Agraria Sacchi Maria Rosa I-TURANO LIDIGIANO (MI)	B2	Full scale (300 m^3)	Completely mixed Int. loaded	RPA Fontana	RPA Fontana	07.1980

	Address	Map	Scale (Size)	System	Constructor	Monitoring	Start-up
232	Cooperativa San Pio X I-VEDELAGO (TV)	C1	Full scale (930 m^3)	Completely mixed Continuous	D. JACOROSSI Spa, Roma	CNEN-FARE Roma	12.1980
233	Azienda Agricola Malintesa I-VILLANOVA D'ARDA (PC)	B3	Full scale (130 m^3)	Completely mixed	EDAECO SRL Cremona		01.1982
234	Baileboro Co-op IRL-BAILEBORO CO-CAVAN	C6	Pilot scale (11 m^3)	Completely mixed Int. loaded (35 °C)	Owner	AN FORAS TALUNTAIS Co-Wexford	08.1978
235	Farm of M. Sheehan, Bandon IRL - CO-CORK	A8	Full scale (34 m^3)	Anaerobic filter	Owner	Owner	04.1982
236	The Agricultural College Athenry IRL - CO-GALWAY	B7	Pilot scale (3.5 m^3)	Anaerobic filter (35 °C)	Owner + local tradesmen	Microbiology Dept, UCG Galway	04.1982
237	New Inn Piggery East Galway Cooperative IRL - CO-GALWAY	B7	Pilot scale (18 m^3)	Anaerobic filter (28-30 °C)	HANLEY BROTHERS	Microbiology Dept, UCG Galway	09.1977 to 09.1978
238	M. E. Van Ee Amersfoortseweg, 287 NL-GARDEREN	D4	Full scale (462 m^3)	Completely mixed Int. loaded (27-30 °C)	Owner + local tradesmen	IMAG Wageningen	Autumn 1978
239	M. G. M. Harte Rustenburgerweg, 61 NL-HEERHUGOWAARD	B2	Full scale (200 m^3)	Completely mixed Int. loaded	Owner + local tradesmen	IMAG Wageningen	04.1980

	Address	Map	Scale (Size)	System	Constructor	Monitoring	Start-up
240	H. Hobert Koperdijk, 5 NL-KLARENBEEK	D4	Full scale (115 m^3)	Completely mixed Int. loaded (30 °C)	PAQUES BV El Balk	IMAG-RAAD Wageningen	12.1979
241	B.H. Hagemeyer Stobbeplasweg, 2 NL-SLAGHAREN	D3	Full scale (500 m^3)	Two-step Int. loaded	Fa. de Boer Leeuwarden	IMAG Wageningen	10.1981
242	A. Wills Middleton Estate, Long Parish UK-ANDOVER, HANTS	F9	Full scale (77.3 m^3)	Completely mixed	Owner	None	1982
243	Brooksby Agricultural College UK-BROOKSBY (Leicester)	F8	Pilot scale (3 m^3)	Completely mixed Int. loaded (35 °C)	Owner	Leicester Polytechnic	04.1980
244	D. Thompson Fosslands Farm, Old Earswick UK-HUNTINGTON (Yorks)	F6	Full scale (450 m^3)	Completely mixed	KILBEES/ FARROWS Ltd. Windsor Forest	ADAS Farrows Ltd.	01.1981
245	Botany Bay Farm Weston UK-HITCHIN, Herts	F8	Full scale (34 m^3)	Completely mixed Int. loaded (35 °C)	FARM GAS Ltd. + Owner Bishops Castle		01.1978
246	Pittrichie piggery White Rashes UK-NEWMACHAR (Aberdeenshire-Scotland)	F3	Full scale (744 m^3)	Completely mixed Int. loaded (38 °C)	HARVESTORE PRODUCTS Ltd. Eye	Rowett Research Institute, W. of Scotland College of Agriculture Aberdeen	10.1980

	Address	Map	Scale (Size)	System	Constructor	Monitoring	Start-up
247	Elmsfield Farm, UK-CHIPPING NORTON, Oxon	F8	Full scale ($105\ m^3$)	Completely mixed Int. loaded (38 °C)	FARM GAS Ltd. + SCS BIOTECHNO-LOGY Ltd Bishops Castle		10.1981
248	Elmsfield Farm UK-CHIPPING NORTON, Oxon		Pilot scale ($2 \times 1.5\ m^3$)	Completely mixed Int. loaded (38 °C)	FARM GAS Ltd. + SCS BIOTECHNO-LOGY Ltd. Bishops Castle		04.1981
249	Elmsfield Farm UK-CHIPPING NORTON, Oxon		Pilot scale ($2 \times 10\ m^3$)	Anaerobic filter (38 °C)	FARM GAS Ltd. + SCS BIOTECHNO-LOGY Ltd. Bishops Castle		01.1980
250	A. Meyer Station SPF CH-1143 APPLES	B3	Full scale ($2 \times 100\ m^3$)	Completely mixed	HUBER AG Lengnau	None	11.1981
251	N. Wick CH-8717 BENKEN	E3	Full scale ($50\ m^3$)	Completely mixed	INVENTA AG Domat	Milchtechn. Labor. ETH Zürich	10.1979
252	M. Stöckli CH-5623 BOSWIL	D3	Full scale ($50\ m^3$)	Completely mixed	WINKLER/Owner	None	1978
253	Gebr. Odermatt CH-6383 DALLENWIL	D3	Full scale ($70\ m^3$)	Plug-flow	LBA Brugg	None	01.1981
254	A. Minikus CH-9113 DEGERSHEIM	E3	Full scale ($2 \times 80\ m^3$)	Completely mixed	BINOTECH AG Roggwil	None	12.1981
255	A. Minikus CH-9113 DEGERSHEIM	E3	Full scale ($157\ m^3$)	Completely mixed	LBA Brugg	None	09.1981

	Address	Map	Scale (Size)	System	Constructor	Monitoring	Start-up
256	H. Güntensperger CH-9601 DREIEN	E2	Full scale (300 m^3)	Fed-batch (15-20 °C)	Owner	Biogas Projekt FAT-Tanikon	11.1979
257	M.J. Berweger CH-8360 ESCHLIKON	E2	Full scale (40 m^3)	Completely mixed	Owner	None	09.1980
258	E. Dürr CH-8500 FRAUENFELD	E2	Full scale	Completely mixed	SIEBER/ ROTAVER AG Andelfingen	None	
259	P. Züger Bodenwiese CH-8854 GALGENEN	E3	Full scale (2 x 500 m^3)	Completely mixed	INVENTA AG Domat	None	1982
260	A. Pfister CH-8638 GOLDINGEN	E2	Full scale	Fed-batch (15 °C)	FRICK Triesenberg	none	1981
261	UFA-Versuchsbetrieb Bühl CH-5604 HENDSCHIKEN	C3	Full scale (140 m^3)	Partially mixed	MANAHL Bregenz (A)	None	1982
262	J. Rohrer CH-8638 HINTERGOLDINGEN	E2	Full scale (800 + 620 m^3)	Fed-batch (15 °C)	FRICK Triesenberg	None	1981
263	A. Näpflin CH-8574 ILLIGHAUSEN	E2	Full scale (120 m^3)	Partially mixed	LBA Brugg	Biogas Projekt FAT-Tanikon	10.1981
264	E. Huber CH-8722 KALTBRUNN	E3	Full scale (157 m^3)	Completely mixed Int. loaded (36 °C)	LBA + owner Brugg	Milchtechn. Labor. ETH Zürich	06.1981
265	F. Schorri Lindachstr. 50 CH-3038 KIRCHLINDACH	C3	Full scale (80 m^3)	Completely mixed	HUBER AG Lengnau	None	12.1981

	Address	Map	Scale (Size)	System	Constructor	Monitoring	Start-up
266	R. Habegger CH-9652 NEU ST JOHANN	E2	Full scale (90 m^3)	Partially mixed	MANAHL Bregenz (A)	None	1981
267	W. Jung CH-9249 NIEDERSTETTEN	E2	Full scale (70 m^3)	Plug-flow	LBA Brugg	Milchtechn. Labor. ETH Tanikon	02.1981
268	P. Bucher CH-5647 OBERRÜTI	D3	Full scale (80 m^3)	Completely mixed	HUBER AG Lengnau	None	1981
269	P.A. Fragniere Laiterie CH-1428 PROVENCE	B3	Full scale (50 m^3)	Completely mixed	INVENTA AG Domat	None	1981
270	E. Graf CH-9325 ROGGWIL	EF1	Full scale (3 x 180 m^3 1 x 600 m^3)	Completely mixed	BINOTECH AG Roggwil	Biogas Projekt FAT-Tanikon	1967
271	A. Reichmuth CH-6430 SCHWYZ	E3	Full scale (198 m^3)	Completely mixed	LBA Brugg	None	09.1981
272	Eidgenössische Forschungs- anstalt für Betriebswirt- schaft und Land, Land- technik CH-8355 TANIKON		Pilot scale (2 x 5 m^3)	Completely mixed	FAT Tanikon	Biogas Projekt FAT-Tanikon	1981
273	Eidgenössische Forschungs- anstalt für Betriebswirt- schaft und Land, Land- technik CH-8355 TANIKON		Pilot scale (5 or 30 m^3)	Completely mixed or accumulation	FAT Tanikon	Biogas Projekt FAT-Tanikon	1981
274	H.-P. Neukomm CH-8240 THAYNGEN	E1	Full scale (135 m^3)	Completely mixed	LBA Brugg	Biogas Projekt FAT-Tanikon	11.1979

	Address	Map	Scale (Size)	System	Constructor	Monitoring	Start-up
275	H. Wenger Schweinemästerei CH-3549 TRIMSTEIN	C3	Full scale (75 m^3)	Completely mixed	Owner + ROTAVER AG Lützelfluh	None	
276	J. Waltenspühl CH-8856 TUGGEN	E3	Full scale	Partially mixed (15 °C)	FRICK Triesenberg	None	12.1980
277	E. Horisberger CH-8476 UNTERSTAMMHEIM	E2	Full scale (180 m^3)	Fed-batch (12-20 °C)	Owner	Biogas Projekt FAT-Tanikon	1978
278	Ch. Schnyder CH-3178 UTTEWIL	C3	Full scale (98 m^3)	Completely mixed	LBA Brugg	Canton de Fribourg	05.1980
279	A. Karlen CH-1445 VUITEBOEUF	B3	Full scale (200 m^3)	Completely mixed Int. loaded (35 °C)	CHAPALLAZ Vuiteboeuf	Biogas Projekt FAT-Tanikon	1978
280	K. Hopli CH-8585 ZUBEN	E2	Full scale (60 m^3)	Completely mixed	LBA Brugg	None	04.1980
			I.4. POULTRY MANURE				
281	Lhermitte Chemin du Tombeu, 4 B-5413 BAILLONVILLE		Full scale (45 m^3)	Plug-flow Int. loaded (35 °C)	AIDR Bruxelles	AIDR Bruxelles	03.1981
282	Van der Locht Kleine Achterstraat, 134 B-3573 WIJCHMAAL (PEER)	E2	Full scale (100 m^3)		EKW PVBA Eksel		1983
283	Szabo D-7967 BAD WALDSEE- TANNWEILER	C9	Full scale (120 m^3)	Completely mixed (32 °C)	WLZ	None	1983

	Address	Map	Scale (Size)	System	Constructor	Monitoring	Start-up
284	G. Hölzl Ihlendorf, 3 D-2845 DAMME 1 (Kreis Vechta)	B34	Full scale (100 m^3)	Completely mixed	ÖKOTHERM Hagen	FAL Braunschweig	11.1982
285	W. Möhler, Muthof D-7119 FORCHTENBERG	C7	Full scale	Plug-flow	FIMA Obersontheim	Universität Hohenheim Stuttgart	01.1982
286	Nordsaatgutsverwaltung Salzau Herr von Zitzewitz D-2301 SALZAU	D2	Full scale (100 m^3)	Completely mixed (35 °C)	ÖKOTHERM Hagen	Institut für Landwirt- schaftliche Verfahrens- technik (Kiel)	12.1981
287	K. Hofmeier Viehbrunnenweg, 3 D-7730 TANNHEIM bei VILLINGEN	B8	Full scale (150 m^3)	Completely mixed	GAE	Landtechnik Weihenstephan	1982
288	Bizzari Graincourt F-57520 DELME	G2	Full scale (2 x 120 m^3)	Continuous	MARUMO Guebwiller		1982
289	M. Fink GAEC "Mont des Saints" F-HEILIGENBERG (Bas Rhin)	H3	Full scale (150 m^3)	Completely mixed Semi-continuous (36 °C)	AGRAGAZ FERMENTECHNIK (CH) + MARUMO Guebwiller	TIMAG IRCHA	06.1980
290	Fischer F-57 LOUVIGNY	G2			METHANECO Montmelian		

	Address	Map	Scale (Size)	System	Constructor	Monitoring	Start-up
			I.5. RABBIT MANURE				
291	E. Deome Rue du jardinet, 42 B-6736 NIVELET L'EGLISE	E4	Full scale (100 m^3)	Plug-flow Int. loaded	AIDR Bruxelles	AIDR Bruxelles	1983
			I.6. DAIRY CATTLE MANURE BEDDING				
292	Domaine expérimental de l'ENSIAA. La Bouzule F-54370 CHAMPENOUX (Nancy)	G3	Pilot scale (15 + 6 m^3)	Two-step (30-35 °C)	INDUFERMA Paris	ENSIAA Nancy	01.1981
293	Chevalier La grande Genière-Javene F-35300 FOUGERES	C3	Full scale (4 x 15 m^3)	Batch	Owner	None	1975
294	GAEC Besson, Ferme de Sedeyrac F-15000 NAUCELLES (CANTAL)	E6	Pilot scale (38 m^3)	Anaerobic filter (37 °C)	Ets COSSOUL AGRINERGIE Aurillac	COSTIC St-Remy-les-Chevreuse	07.1981
295	GAEC Besson Ferme de Sedeyrac F-15000 NAUCELLES (CANTAL)	E6	Pilot scale (41 m^3)	Completely mixed Int. loaded (37 °C)	Ets COSSOUL AGRINERGIE Aurillac	COSTIC St-Remy-les-Chevreuse	06.1981
296	Domaine de Borret F-31370 POUCHARRAMET (Hte Garonne)	D7	Pilot scale (5 m^3)	Plug-flow Int. loaded (35 °C)	ELF-BIO RECHERCHE + INSA	INSA Toulouse	06.1981
297	R. Sandot Evette Salbert F-90300 VALDOIE	G3	Full scale (5 + 7 m^3)	Batch	Owner	None	1973
298	A. Maillet F-74290 VEYRIER-DU-LAC	G5	Full scale (130 m^3)	Completely mixed Int. loaded (35-37 °C)	Sté BATRIGAL	None	End of 1980

	Address	Map	Scale (Size)	System	Constructor	Monitoring	Start-up
299	"Les amis de l'homme" Val de Saint Germain F-47300 VILLENEUVE-SUR-LOT	D6	Full scale (4 x 22 m^3)	Batch (20-25 °C)	Owner	None	1976
300	A. Chapuis Le moulin Maurt F-88200 XERTIGNY	G3	Full scale (3 x 8 m^3)	Batch	Owner	None	1952
			I.7. BOVINE CATTLE MANURE BEDDING				
301	R. Kring, Landgraf Friedrich str., 40 D-FRANKENBERG/ FRIEDRICHSHAUSEN	C5	Full scale (100 m^3)	Plug-flow (33 °C)	Owner	Gesamthoch-schule Kassel + KTBL Darmstadt	09.1981
302	Gutsverwaltung Erlacher Höhe D-7151 GROSSERLACH	C7	Full scale (200 m^3)	Plug-flow (35 °C)	LIPP GmbH Tannhausen	None	04.1982
303	D. Reusch Marktstr., 4 D-7425 HOHENSTEIN /BERNLOCH	C8	Full scale (21 m^3)	Plug-flow (35 °C)	Owner	None	01.1981
304	A. Hau Austrasse, 2 D-8150 HOLZKIRCHEN	E9	Full scale (15 m^3)	Completely mixed (38 °C)	Owner	None	12.1982
305	D. Müller Buschweg, 14 D-2358 KATTENDORF	CD2	Full scale (150 m^3)	Batch (35 °C)	Owner	Inst. f. Landwirt-schaftliche Verfahrens-technik, Kiel	12.1982

	Address	Map	Scale (Size)	System	Constructor	Monitoring	Start-up
306	Abtei Münsterschwarzach Schweinfurter Str. 40 D-8711 MUNSTERSCHWARZACH	D67	Full scale (250 m^3)	Completely mixed (33 °C)	Gesellschaft für Alternative Energierzeugung mbH Michelfeld	Landtechnik Weihenstephan	12.1982
307	F. Bross D-7164 OBERFISCHACH	C7	Full scale (40 m^3)	Plug-flow	FIMA Obersontheim	None	1981
308	Labeyrie F-40800 AIRE-SUR-ADOUR	C7	Full scale	Batch	SETAL Pau	IRCHA	End of 1982
309	Leroux, La salle F-53420 CHAILLAND	C3	Full scale (2 x 11 + 1 x 17 m^3)	Batch	Owner	None	08.1980
310	Baudry La Fiere - la Suguinière F-49300 CHOLET	C4	Full scale (2 x 20 m^3)	Batch	Sté SEBM Lambersart		1982-1983
311	Ledonne, La Frogerie St Ouen Le Brisout F-61410 COUTERNE	D3	Full scale (4 x 20 m^3)	Batch	Owner		1982-1983
312	Station expérimentale de Derval, La Touche F-44590 DERVAL	C4	Full scale	Batch			
313	Jade, Cosquer Beuzec Cap Sizun F-29100 DOUARNENEZ	A3	Full scale (2 x 12 m^3)	Batch	Owner		

	Address	Map	Scale (Size)	System	Constructor	Monitoring	Start-up
314	Marché aux Bestiaux Parc de l'Aumaillerie F-35300 LA-SELLE-EN-LUITRE	C3		Batch			
315	Hameau de la Bizotière F-50190 MARCHESIEUX (Manche)	C2	Full scale (47 m^3)	Batch	Local tradesmen	ANERA + INRA Caen	03.1982
316	Defaye, Froment Saint-Privat des Prés F-24410 ST-AULAYE	D6	Full scale (9 m^3)	Batch	Owner		1981
317	Mida-Tra, Pépieux Castelnau - Barbarens F-32450 SARAMON	D7	Full scale (4 x 300 + 1 x 10 m^3)	Batch (4 x) Int. loaded (1 x)	MIDA-TRA Saramon	MIDA-TRA Saramon	1980
318	Maudet La Gicquelaie F-56460 SERENT	B34		Batch			
319	Lycée agricole F-41100 VENDOME	D3	Pilot scale (3 x 15 m^3)	Batch	Local tradesmen	Lycée agricole de Vendôme	03.1982
320	Bourdon Kermenez Le Juch F-29 (Finistere)	A3		Batch			
321	Kerouedan Pont-Croix F-29 (Finistère)	A3		Batch			
322	Motin St-Clément les Places F-69000 (Rhone)	F5	Full scale (2 x 20 m^3)	Batch	METHANECO Montmelian		

	Address	Map	Scale (Size)	System	Constructor	Monitoring	Start-up
323	Ridge Farm, Linley UK-BISHOPS CASTLE, Salop	E7	Pilot scale (11 m^3)	Completely mixed Int. loaded (35 °C)	FARM GAS Ltd. Bishops Castle	Farm Gas Ltd. Bishops Castle	1976
			I.8. PIG MANURE BEDDING				
324	W. Freitag, Fritzen- mühle D-3549 TWITZETAL- BERNDORF	C5	Full scale (30 m^3)	Plug-flow (35 °C)	Owner	Gesamthoch- schule Kassel + KTBL Darmstadt	01.1982
325	M. Holste Egenbostel, 9 D-2722 VISSELHOVEDE	C3	Full scale (32 m^3)	Plug-flow (34 °C)	Owner	FAL Braunschweig	01.1982
326	J. Leriche, Le Nelie- marcilly d'Azergues F-69300 LOZANNE (Rhône)	F5	Full scale (2 x 18 m^3)	Batch (ambient T)	Owner + local tradesmen	None	1956
			I.9. POULTRY MANURE BEDDING				
327	Petre Rue St-Anne, 2 B-5918 HELECINE	D2	Full scale (2 x 300 m^3)	Plug-flow Int. loaded	AIDR Bruxelles	AIDR Bruxelles	1983
328	Sté Bretonne de découpe de dinde - zone indus- trielle, Pleucadec F-56140 MALESTROIT	B3	Full scale (2 x 25 m^3)	Completely mixed Int. loaded (35 °C)	Local tradesmen	GUYOMARC'H Vannes	10.1980
329	Le Picard Kermoysan F-56500 MOREAC	B4	Full scale (300 m^3)	Batch	Owner	Coop. Agri- culteurs du Morbihan	1980

	Address	Map	Scale (Size)	System	Constructor	Monitoring	Start-up
330	Tisseau, La Davière F-49310 ST-PAUL-DES- BOIS	C4	Full scale	Batch			1980
			I.10. RABBIT MANURE BEDDING				
331	Cozot , Villemandeur F-45700 ST-MAURICE-SUR-FESSARD	E4	Full scale	Batch	Owner		1976
			I.11. GOAT MANURE BEDDING				
332	A. Eylieu Saint Ferreol F-84500 BOLLENE	F6	Full scale (2 x 23 m^3)	Batch (35 °C)	METHANECO Montmelian	None	02.1980
			I.12. SHEEP MANURE BEDDING				
333	L. Montheil La Peinetre-Paunat F-24510 ST-ALVERE	D6	Full scale (12 m^3)	Batch	Owner	None	1974
			I.13. HORSE MANURE BEDDING				
334	Haras des Bréviaires F-78610 LE PERRAY-EN-YVELINES	E2	Pilot scale (34 m^3)	Batch	Owner	None	11.1980
			I.14. STRAW				
335	K. Bohnenberger D-8170 OBERFISCHBACH (bei Bad Tölz)	E9	Full scale (38 m^3)	Plug-flow	Landwirtschaftlicher Anlagenbau (W. Schraufstetter) Ismaning	None	12.1982

No.	Address	Map	Scale (Size)	System	Constructor	Monitoring	Start-up
			I.15. CROP RESIDUES				
336	Firma Zegwaard Curacaostraat, 1 NL-DELFT	B5	Pilot scale (10 m^3)	Upflow without carrier Two-step (30-32 °C)	SPAANS HOOFD-DORP + local tradesmen	IBVL Wageningen	10.1981
			I.16. DAIRY CATTLE + PIG MANURE				
337	Gebr. Buysse L. en E. Beekstraat, 55 B-9968 BASSEVELDE	B1	Full scale (150 m^3)	Plug-flow	IVPO Mariakerke	IVPO/RUG Gent	1983
338	Fiskebaekgaard Graasten Landbrugsskole DK-6300 GRAASTEN	C6	Full scale (2 x 220 m^3)	Completely mixed (2 x in series) Int. loaded (35 $\pm$ 1 °C)	SYDJYSK/AGRO-ANLAEG + AF. PETESENS MASKINFABRIEK	STUB Glostrup	09.1979
339	Sj. F. Bygholen DK-8700 HORSENS	C4	Full scale (130 m^3)	Completely mixed	Local tradesmen + Technological Institute	Carl Bro Ltd Glostrup	08.1982
340	K. Kirk Enkerbro, 9 DK-7752 SNEDSTED	A3	Full scale (4 x 200 m^3)	Batch (35 °C)	Owner + local tradesmen	STUB Glostrup	06.1981
341	Ferme de l'EDE des côtes du Nord F-GRECOM	B3	Full scale	Anaerobic filter (35 °C)	AIR LIQUIDE Champigny sur Marne	Air Liquide Champigny sur Marne	1983

	Address	Map	Scale (Size)	System	Constructor	Monitoring	Start-up
342	H.J. Meerdink Hoeninkdijk, 4 NL-BREDEVOORT	E4	Full scale (660 m^3)	Completely mixed Int. loaded (31 °C)	Local tradesmen	IMAG Wageningen	07.1981
343	P. Kuunders Weyer, 6 NL-MILHEEZE	D5	Full scale (330 m^3)	Completely mixed Int. loaded	PAQUES BV El Balk	IMAG Wageningen	01.1982
344	De Leiding 6 NL-WITHAREN	D3	Full scale (195 m^3)	Completely mixed Int. loaded (30 °C)	PAQUES BV El Balk	IMAG Wageningen	01.1980
345	Walnut Tree Farm Chediston UK-HALESWORTH, Suffolk	G7	Full scale (360 m^3)	Completely mixed Int. loaded (35 °C)	HAMWORTHY ENGINEERING Ltd. /CHEDISTON- AGRI SYSTEMS Poole	Cardiff Uni- versity + Ham- worthy Engi- neering	08.1980
346	Chapel Farm, Ringshall UK-STOWMARKET, Suffolk	G8	Full scale (340 m^3)	Completely mixed Int. loaded (32 °C)	FARM GAS Ltd. Bishops Castle		04.1978
347	A. Nietlisbach CH-5637 BEINWIL	D3	Full scale (80 m^3)	Completely mixed	HUBER AG Lengnau	None	10.1980
348	Kaufmann CH-9442 BERNECK	F2	Full scale	Completely mixed	LBA Brugg	None	1981
349	F. Locher CH-8906 BONNSTETTEN	D2	Full scale (50 m^3)	Completely mixed (33 °C)	LBA Brugg	Biogas Projekt FAT-Tanikon	12.1978

	Address	Map	Scale (Size)	System	Constructor	Monitoring	Start-up
350	J. Köchli CH-5630 BUTTWIL	D3	Full scale (60 m^3)	Completely mixed	LBA Brugg	None	08.1981
351	H. Meier Untergris CH-8546 ELLIKON	E2	Full scale (80 m^3)	Completely mixed	ROTAVER AG Lützelfluh	None	
352	K. Künzle CH-8734 ERMENSWIL	E3	Full scale (2 x 70 m^3)	Plug-flow	Owner	None	1980
353	R. Stämpfli, Feld CH-3066 FERENBERG	C3	Full scale (60 m^3)	Completely mixed	HUBER AG Lengnau	None	1982
354	O. Schnider CH-6173 FLÜHLI	D3	Full scale (100 m^3)	Completely mixed	SCHWAB/owner Kersers	None	02.1980
355	Kloster Frauenthal CH-6331 HAGENDORN	D3	Full scale (285 m^3)	Completely mixed Int. loaded (30 °C)	LBA Brugg	Biogas Projekt FAT-Tanikon	06.1979
356	N. Eisenring CH-9243 JONSCHWIL	E2	Full scale (70 m^3)	Plug-flow	LBA Brugg	None	09.1980
357	H. Luder CH-3422 KIRCHBERG	C3	Full scale (100 m^3)	Completely mixed	HUBER AG Lengnau	None	1981
358	B. Ballmer CH-3045 MEIKIRCH	C3	Full scale (2 x 150 m^3)	Fed-batch	LBA Brugg	None	08.1981
359	R. Weber CH-8932 METTMENSTETTEN	D3	Full scale (80 m^3)	Completely mixed	HUBER AG Lengnau	None	1981

	Address	Map	Scale (Size)	System	Constructor	Monitoring	Start-up
360	Landwirtschaftliche Schule Mangettes CH-1870 MONTHEY	F2	Full scale (2 x 100 m^3)	Completely mixed	HUBER AG Lengnau	Owner	1982
361	W. Kripl, Schloss Watt CH-9402 MÖRSCHWIL	E2	Full scale (2 x 80 m^3)	Completely mixed	BINOTECH AG Roggwil	None	12.1981
362	Landwirtschaftliche Schule Schwand CH-3110 MUNSINGEN	C3	Full scale (157 m^3)	Completely mixed	LBA Brugg	None	01.1981
363	A. Zürcher CH-3431 RANFLUH	C3	Full scale (60 m^3)	Completely mixed	HUBER AG Lengnau	None	09.1980
364	J. Gloggner CH-6017 RUSWIL	E3	Full scale (106 m^3)	Completely mixed Int. loaded	Owner	None	1980
365	J. Wyser CH-6017 RUSWIL	E3	Full scale (155 m^3)	Completely mixed	HUBER AG Lengnau	None	10.1979
366	F. Schneider CH-9464 RÜTHI	F2	Full scale (98 m^3)	Completely mixed	LBA Brugg	None	1981
367	Landwirtschaftliche Schule Rheinhof CH-9465 SALEZ	F2	Full scale (100 m^3)	Completely mixed	HUBER AG Lengnau	Owner	1982
368	Handelsmühle CH-5040 SCHÖFTLAND	D2	Full scale (60 m^3)	Completely mixed	HUBER AG Lengnau	None	1981
369	F. Roth CH-2615 SONVILIER	B2	Full scale (50 m^3)	Completely mixed	LBA Brugg	Biogas Projekt FAT-Tanikon	03.1979

	Address	Map	Scale (Size)	System	Constructor	Monitoring	Start-up
370	Strafanstalt CH-6243 WAUWILERMOOS	D3	Full scale (180 m^3)	Partially mixed	MANAHL Bregenz	None	1982
371	Strafanstalt CH-3236 WITZWIL	B3	Full scale (340 m^3)	Completely mixed	LBA Brugg	None	12.1981
372	H. Wittwer CH-3052 ZOLLIKOFEN	C3	Full scale (65 m^3)	Completely mixed	LBA Brugg	None	08.1981
			I.17. <u>BOVINE CATTLE + PIG MANURE</u>				
373	Ferme expérimentale Sart Tilman Université de Liège B-4000 LIEGE		Full scale ($\pm$ 300 m^3)	Completely mixed	Local tradesmen	Institut de Pathologie Univ. Liège Sart Tilman	1983
374	A. Holzwimmer Wipfelsberg, 108 D-8261 REISCHACH	E8	Full scale (140 m^3)	Plug-flow (30 °C)	BVT Bregenz (A)	Landwirt- schaftskammer Altötting	01.1981
375	E. Döpper, Prickingshof Lehmbraken, 14 D-4358 HALTERN-SYTHEN	B4	Full scale (1300 m^3)	Completely mixed (20 °C)	STEINMANN + ITTIG Minden	None	07.1982
376	Westkamper Sennelagerstr. 69 D-4790 SANDE bei PADERBORN	C4	Full scale (80 m^3)	Completely mixed (32 °C)	Owner	FAL + Braun- schweig Univ. Oldenburg	1980
377	F. Gartelmann Neuenkirchen, Im Klint 11 D-2822 SCHWANEWEDE	C3	Full scale (96 m^3)	Plug-flow (34 °C)	LIPP GmbH Tannhausen	None	11.1981

	Address	Map	Scale (Size)	System	Constructor	Monitoring	Start-up
378	Avlsgarden Saxenhoej kalkbraenderivej DK-4990 SAKSKØBING	E7	Full scale (170 m^3)	Completely mixed Int. loaded	Local tradesmen		06.1981
379	L. Bruhn, Højen Oxviggaard DK-7100 VEJLE	B5	Full scale (100 m^3)	Completely mixed Int. loaded (36-37 °C)	HØRLYCH & KONGSTED Vamdrup		10.1979
380	Coupe, GAEC de la Villepoissin Henanbihen F-22550 MATIGNON	B3		Continuous			
381	Stalla Sociale COAGRI via Levata I-NOVELLARA (RE)	C3	Full scale (1200 + 500 m^3)	Two-step Completely mixed Int. loaded (32 °C)	TERMOMECCANICA La Spezia	CRPA FARE Reggio Emilia	1982
			I.18. DAIRY CATTLE + POULTRY MANURE				
382	American Farm School GR-THESSALONIKI	D2	Full scale (300 m^3)	Int. loaded	American Co. + local tradesmen		1983
			I.19. BOVINE CATTLE + POULTRY MANURE				
383	J. Vermelfoort Balledonk, 4a NL-5473 BE HEESWIJK	C5	Full scale (320 m^3)	Completely mixed Int. loaded	Fa. v. de MEIJDEN Heeswijk	IMAG Wageningen	01.1982
			I.20. PIG + POULTRY MANURE				
384	W. Riezebosch Wencopperweg, 36 NL-BARNEVELD	C4	Full scale (145 m^3)	Completely mixed Int. loaded (27-30 °C)	PAQUES BV El Balk	IMAG Wageningen	11.1980

	Address	Map	Scale (Size)	System	Constructor	Monitoring	Start-up
385	Van Gaal Molendijk, 63 NL-SCHIJNDEL	C5	Full scale (280 m^3)	Completely mixed Int. loaded (26-29 °C)	VAN WANROOY Heeswijk	IMAG Wageningen	09.1980
386	Cannes Farm, Mansons Cross UK-MONMOUTH, Gwent	E8	Full scale (330 m^3)	Completely mixed Int. loaded (33 °C)	Owner	None	10.1980
			I.21. PIG MANURE + CHEESE DAIRY WASTEWATER				
387	Azienda Agricola S. Matteo I-TRAVERSETOLO (PR)	C3	Full scale (350 m^3)	Completely mixed Int. loaded (32-35 °C)	SIDI Parma	CRPA- CNEN FARE Reggio Emilia	1982
388	H. Allemann CH-6403 KUSSNACHT	E3	Full scale	Fed-batch (± 15 °C)	FRICK Triesenberg	None	10.1980
			I.22. DAIRY CATTLE MANURE + CROP RESIDUES				
389	W. Evans Caerfai Farm, St. Davids UK-HAVERFORDWEST, Pembs	D8	Full scale (71 m^3)	Plug-flow (30 °C)	Owner	None	11.1979
			I.23. PIG MANURE + CROP RESIDUES				
390	Cleppa Park, COED Kermew Nr Newport UK-GWENT	E8	Pilot scale (1.5 m^3)	Partially mixed (hydraulic) (35 °C)	Owner + CAM- BRIAN PLASTICS Ltd.	Cardiff University	08.1980
391	Cleppa Park, COED Kermew Nr Newport UK-GWENT	E8	Pilot scale (1.5 m^3)	Anaerobic filter (35 °C)	Owner	Cardiff University	10.1981

	Address	Map	Scale (Size)	System	Constructor	Monitoring	Start-up
392	Cleppa Park, COED Kermew Nr Newport UK-GWENT	E8	Pilot scale (30 m^3)	Plug-flow (ambient T)	Owner + CAM-BRIAN PLAS-TICS Ltd.	Cardiff University	03.1981
393	Cleppa Park, COED Kermew Nr Newport UK-GWENT	E8	Pilot scale (30 m^3)	Anaerobic contact (35 °C)	HAMWORTHY ENGINEERING Ltd., Poole	Cardiff University	01.1982
	I.24. POULTRY MANURE + OLIVE OIL WASTEWATERS + OTHER WASTES						
394	Agricultural College of Athens, Iera Odos, 75 Botanikos GR-ATHENS		Pilot scale (5 m^3)	Plug-flow (35 °C)	Owner	Agricultural College of Athens	05.1982
	I.25. DAIRY CATTLE + HORSE MANURE						
395	Azienda La Margherita I-CARMAGNOLA (TO)	A2	Full scale (70 m^3)	Completely mixed	FM ORAENERGIA Cacciola		07.1981
	I.26. DAIRY CATTLE + WASTEWATERS OF SLAUGHTER-HOUSE						
396	Azienda Agricola Goglia Calabrese I-VITULANO (BN)	E5	Full scale (500 m^3)	Completely mixed	SEA Srl		07.1982
	I.27. PIG MANURE + DAIRY WASTEWATERS (+ SLUDGES)						
397	Caseificio Sociale I-SORANO (GR)	C4	Full scale (2 x 900 m^3)	Completely mixed Continuous	ACTA Roma		09.1982
	I.28. HORSE MANURE + GARDEN RESIDUES						
398	M. Demont CH-1253 VANDOEUVRES	A4	Full scale (12 m^3)	Plug-flow	PFULG Cologny	None	02.1980

	Address	Map	Scale (Size)	System	Constructor	Monitoring	Start-up
	I.29. DAIRY CATTLE + PIG + POULTRY MANURE						
399	W. F. J. Roozen Beerzeweg, 12 NL-DIESSEN	C6	Full scale (400 m^3)	Plug-flow Int. loaded	Fa. VAN MANROOY Heeswijk	IMAG Wageningen	03.1982
400	J. Barmettler CH-5746 WALTERSWIL	C3	Full scale (3 x 100 m^3)	Completely mixed	HUBER AG Lengnau	None	1982
401	F. Siegenthaler CH-8116 WURENLOS	D2	Full scale (100 m^3)	Completely mixed	HUBER AG Lengnau	None	1981
	I.30. MANURE + CROP RESIDUES + WASTEWATERS + SEPTIC TANK						
402	National Centre for Alternative Technology, Machyulleth UK-POWYS, Wales	D7	Pilot scale (1.9 m^3)	Int. loaded (8-31 °C)	FARM GAS Ltd. Bishops Castle		08.1979
403	F. Jeanneret CH-2534 ORVIN	B3	Full scale (3 m^3)	Completely mixed	WAHL Zollikon	None	1980
	I.31. POULTRY MANURE BEDDING + BOVINE CATTLE MANURE BEDDING						
404	J. Frison Chemin du Bonlas B-7623 GIBECQ	C3	Full scale (95 m^3)	Plug-flow Int. loaded	BIOMETHANE Bruxelles	Centre Agro. de Rech. App. du Hainaut-ATH CARAH	09.1982
405	H. Esslinger Bernauer Strasse D-7922 HERBRECHTINGEN	D8	Full scale (60 m^3)	Plug-flow (38 °C)	LIPP GmbH Tannhausen	None	10.1980

	Address	Map	Scale (Size)	System	Constructor	Monitoring	Start-up
	I.32. PIG MANURE BEDDING + BOVINE CATTLE MANURE BEDDING						
406	H. Schröder Aarweg, 163 D-3559 LICHTENFELS-GODDELSHEIM	C5	Full scale (25 m^3)	Plug-flow (33 °C)	Owner	Gesamthoch-schule Kassel + KTBL Darmstadt	12.1982
407	H. Hemme Markstr. 2 D-3002 WEDEMARK 1	C34	Full scale (60 m^3)	Plug-flow (33-38 °C)	ÖKOPLAN Braunschweig	FAL + Inst. für Siedlungs-wasserwirt-schaft und Abfalltechnik Univ. Hannover	06.1981
408	M. Raimbault, GAEC des Grimaudières - Livre la Touche F-53400 CRAON	C3	Full scale (3 x 25 m^3)	Batch (35-36 °C)	Owner + local tradesmen	None	06.1980
	I.33. DAIRY CATTLE MANURE BEDDING + SHEEP MANURE BEDDING						
409	R. Lafond, Comps F-26220 DIEULEFIT	FG6	Full scale (2 x 33 m^3)	Batch	METHANECO Montmelian	None	1980
	I.34. BOVINE CATTLE MANURE BEDDING + SHEEP MANURE BEDDING						
410	Cesbron, Chemerard-Somoire F-49360 MAULEVRIER	C4	Full scale (5 x 20 m^3)	Batch (20-22 °C)	Owner	None	06.1980
411	Morille, Lavau F-49340 TREMENTINES	C4	Full scale (8 x 10 m^3)	Batch (20 °C)	Owner	None	06.1980

	Address	Map	Scale (Size)	System	Constructor	Monitoring	Start-up
	I.35. SHEEP MANURE BEDDING + GOAT MANURE BEDDING						
412	Cuma des Sources La Batie des Fonds F-26000 LUN-EN-DIOIS	G6	Pilot scale	Int. loaded	Owner + local tradesmen	INRA + IRCHA Montpellier	1982-83
	I.36. VARIOUS MANURE BEDDING						
413	Station expérimentale de l'Institut Technique des Céréales et des Fourrages F-91720 BOIGNEVILLE	E3	Pilot scale (10 + 15 m^3)	Batch	Owner + local tradesmen	ITCF Vert-le-Petit	1976
	I.37. TYPE OF AGRICULTURAL WASTE UNKNOWN						
414	LIPP GmbH (movable plant) D-7091 TANNHAUSEN OSTALBKREIS	D7	Full scale (35 m^3)	Plug-flow	LIPP GmbH Tannhausen	None	1980
415	Lycée d'Yvetot F-76190 YVETOT	D2			Sté SIGOURE + owner Roanne		1982-83
416	Tuloup Vernazoubre - Diovalquière F-34650 LUNAS	E7	Pilot scale (20 m^3)		Owner	INRA + IRAT Montpellier	1982-83
417	F. Rechsteiner CH-8586 ERLEN	E2	Full scale	Completely mixed	BUFFONI	None	1981
418	Eidgenössische Forschungs-anstalt für Agrikultur-chemie Umwelthygiene CH-3097 LIEBEFELD		Pilot scale (2 x 3 m^3)	Batch	FAU Liebefeld	FAU Liebefeld	1979

	Address	Map	Scale (Size)	System	Constructor	Monitoring	Start-up
419	A. Forestier CH-1065 THIERRENS	C3	Full scale	Completely mixed	ROTHEN Moudon	None	1981
420	E. Sprenger, Anetswil CH-9545 WÄNGI	E2	Full scale (2 x 100 m^3)	Plug-flow	PRINZING	None	1982
			II. <u>ENERGY CROPS</u>				
			II.1. <u>Algae</u>				
421	Azienda Agricola "Palazzo" I-LAMEZIA-TERME		Pilot scale (1 m^3)	Completely mixed ($\pm$ 35 °C)	Belgian tradesmen	Istituto di Microbiologia Agraria Tecnica Firenze	10.05.1982

III. DOMESTIC RESIDUES

	Address	Map	Scale (Size)	System	Constructor	Monitoring	Start-up
422	Deponie Ahrenshöft D-2251 AHRENSHÖFT b. HUSUM (Landkr. Nordfriesl.)	C1	Full scale (8.5 ha)	Landfill (horizontal drainage + 10 vertical probes)	MÜLL-EX GmbH West u. Co. KG.	None	10.1981
423	Deponie "Grube-Weiss", D-507 BERGISCH-GLADBACH	AB5	Full scale (1.2 ha)	Landfill (vertical gas pipelines)	INTERATOM Bergisch	None	07.1981
424	Städtische Mülldeponie Braunschweig Ortsteil Watenbüttel D-3300 BRAUNSCHWEIG	D4	Full scale (3.5 ha)	Landfill (vertical gas pipelines)	Dr Ing. Zander Braunschweig	Inst. für Stadtbauwesen Abtlg. Siedlungwasserwirtschaft Universität Braunschweig	09.1982
425	Deponie Datteln-Löringshoff D-4354 DATTELN	A5	Full scale (3.0 ha)	Landfill (horizontal gas pipelines)	Abfallbeseitigungsbetriebe, Ruhrgebiet GmbH I.G. -Essen	None	02.1980
426	Zentraldeponie Emscherbruch D-465 GELSENKIRCHEN	A5	Full scale	Landfill (horizontal and vertical pipelines)	Abfallbeseitigungsbetriebe, Ruhrgebiet GmbH I.G. - Essen	None	
427	Sondermülldeponie Gerolsheim D-6711 GEROLSHEIM	B7	Full scale (15 ha)	Landfill (vertical probes)	Gesellschaft zur Beseitigung Von Sonderabfällen in Rheinland/Pfalz mbH Frankenthal	None	05.1981

	Address	Map	Scale (Size)	System	Constructor	Monitoring	Start-up
428	Deponie Hannover D-3000 HANNOVER	CD4	Full scale	Landfill (horizontal gas pipes)	Fuhramt der Stadt Hannover	None	
429	Deponie Moschheim D-5431 MOSCHHEIM	C4	Full scale (8.4 ha)	Landfill (vertical probes)	Westerwaldkreis Abfallbeseitigung Moschheim	None	
430	Deponie Hohberg D-7530 PFORZHEIM	C7	Full scale (10 ha)	Landfill	Tiefbauamt u. Stadtwerke der Stadt Pforzheim	Inst. für Siedlungswasserbau, Wassergüte und Abfallwirtschaft, Stuttgart	03.1981
431	Deponie "Am Lemberg" b. Poppenweiler, Landkreis Ludwigsburg D-7000 STUTTGART	C78	Full scale (16 ha)	Landfill (vertical pipelines + horizontal drainage)	Büro für Kies u. Abfall W. Ryser Uttingen Switzerland	Inst. für Siedlungswasserbau, Wassergüte und Abfallwirtschaft, Stuttgart	11.1981
432	Décharge d'Arnouville-les-Mantes F-78790 SEPTEUIL	D2	Full scale (2.5 ha)	Landfill		Sté. France - Déchets	1982
433	Décharge F- BARLIN, BEUVRY, FRETIN (Nord)	E1	Pilot scale	Landfill		ANRED, GDF, INRA + Service géologique Nord-Pas de Calais du BRGM	1983

	Address	Map	Scale (Size)	System	Constructor	Monitoring	Start-up
434	Décharge F - ROCHE-LA-MOLIERE (Loire)	F5	Pilot scale	Landfill		Ecole des mines de St-Etienne	1982
435	CARENE, Le mas de Vinay 13, rue de l'Argenterie F-38360 SASSENAGE	G6	Pilot scale	Digester			1982
436	Décharge de Montaubert F-91820 VERT-LE-GRAND (Essonne)	E3	Full scale (9 ha)	Landfill		ANRED-GDF Levallois	1982
437	A.M.R.R. I-TORINO	A2	Full scale (15 ha)	Landfill	AMRR Torino		1981
438	NL-BARNEVELD		Pilot scale	Landfill	GRONTMIJ NV Bilthoven		1983
439	NL-BAVEL	C6	Full scale	Landfill	GRONTMIJ NV Bilthoven		1982
440	NL-DELDEN	E3	Full scale	Landfill	HEIDEMIJ/IVA Arnhem		1982
441	NL-MARKELO		Pilot scale	Landfill	GRONTMIJ NV Bilthoven		1983
442	NL-NUNEN		Pilot scale	Landfill	RAGOB		1983
443	NL-OSS	C5	Full scale	Landfill	GRONTMIJ NV Bilthoven		1983

	Address	Map	Scale (Size)	System	Constructor	Monitoring	Start-up
444	NL-VEENDAM		Pilot scale	Landfill	GRONTMIJ NV Bilthoven		1983
445	NL-WESTWOUD		Pilot scale	Landfill	GRONTMIJ NV Bilthoven		1983
446	NL-WIJSTER		Pilot scale	Landfill	VAM, Wijster		1980
447	NL-WINTERSWIJK		Pilot scale	Landfill	GRONTMIJ NV Bilthoven		1983
448	NL-ZUTPHEN		Pilot scale	Landfill	GRONTMIJ NV Bilthoven		1983
449	Sandy Lane Landfill Site UK-AVELEY, Essex G.L.C.	G8	Full scale (20 ha)	Landfill			07.1983
450	Stairfoot landfill UK-BRANSLEY, Yorks	EF6	Pilot scale (1.2 ha)	Landfill			1981-82 (6 weeks)
451	UK-BLETCHLEY	F8	Full scale	Landfill	LBC Stewartby		
452	UK-BROGBOROUGH	F8	Full scale	Landfill	LBC Stewartby		
453	UK-CALVERT, Bucks	F8	Full scale	Landfill	LBC Stewartby		
454	UK-NORMANTON Brick Co	E7	Full scale (2.8 ha)	Landfill			
455	Wakefield Rd. UK-NORMANTON, Yorks		Pilot scale (1 ha)	Landfill			1982

	Address	Map	Scale (Size)	System	Constructor	Monitoring	Start-up
456	L - Field, London Brick Landfill Ltd. UK-STEWARTBY, Bedford	F8	Full scale (72 ha)	Landfill (20-30 °C)	LBC Stewartby	LBA + ETSU + Dept. of Public Health Engineering London	1979
457	UK-YORKSHIRE Brick Co.	EF6	Full scale (12 ha)	Landfill			1984

IV. INDUSTRIAL WASTES

	Address	Map	Scale (Size)	System	Constructor	Monitoring	Start-up
	IV.1. SUGARY WASTEWATER						
458	Sucrerie de Brugelette B-7940 BRUGELETTE	C3	Full scale (1 x 1660 m^3)	IRIS (37 °C)	Local tradesmen	Sucrerie de Brugelette et Tirlemont	1979
459	Sucrerie de Genappe B-1470 GENAPPE	C3	Full scale (2 x 1530 m^3)	IRIS (37 °C)	Local tradesmen	Sucrerie de Genappe et Tirlemont	1980
460	Sucrerie Naveau B-4250 HOLLOGNE-SUR-GEER		Pilot scale (18 m^3)	Upflow without carrier (32-35 °C)	ACEC DE KONINCKX Charleroi	ACEC + UCL-GEBI (Louvain-la-Neuve) + sucrerie (Hollogne)	10.1981
461	Sucrerie Naveau B-4250 HOLLOGNE-SUR-GEER	D3	Full scale (750 m^3)	Upflow without carrier (32-35 °C)	ACEC DE KONINCKX Charleroi	ACEC + UCL-GEBI (Louvain-la-Neuve) + sucrerie (Hollogne)	09.1982
462	Sucreries Réunies de QUEVY-PERONNES B-7281 QUEVY-LE-GRAND	C3	Full scale (1 x 1530 m^3)	IRIS	Local tradesmen	None	1979-80
463	Zuckerfabrik Brühl AG Postfach 1980 D-5012 BEDBURG	A5	Full scale (1560 m^3)	Upflow without carrier	Braunschweigische Maschinenbauanstalt (BMA) Braunschweig	Inst. für Landw. Technologie u. Zuckerindustrie, Braunschweig	1981 (use of gas)

	Address	Map	Scale (Size)	System	Constructor	Monitoring	Start-up
464	Pfeifer u Langen Postfach 49 D-5350 EUSKIRCHEN	A6	Full scale (5000 m^3)	Anaerobic contact	AB-SORIGONA Staffanstorp (Sweden)	None	1980 (use of gas)
465	Hannoversche Zucker AG, Werk Rethen, Postfach 7 D-3000 HANNOVER-RETHEN	CD4	Full scale (4000 m^3)	Anaerobic contact	Owner	None	1981 (use of gas)
466	Zuckerfabrik Königslutter AG, Postfach 21 D-3308 KONIGSLUTTER	D4	Full scale (1200 m^3)	Upflow without carrier	Gebrüder SULZER Winterthur Switzerland	None	1982 (use of gas)
467	Lehrter Zucker AG Postfach 1160 D-3301 LEHRTE	D4	Full scale (5400 m^3)		Owner	None	1982 (use of gas)
468	Union Zucker Südhannover GmbH, Werk Nordstemmen Postfach 10 D-3204 NORDSTEMMEN	C4	Full scale (16000 m^3)	Anaerobic contact	Owner	None	1980 (use of gas)
469	Zuckerfabrik Franken GmbH Postfach 21 D-8703 OCHSENFURT	CD7	Full scale (1400 m^3)	Upflow without carrier	Braunschwei- gische Maschi- nenbauanstalt (BMA) Braunschweig	FAL Braunschweig	1983 (use of gas)
470	Sucrerie de Vauciennes F-60117 VAUMOISE (Picardie)	E2	Full scale (1600 m^3)	IRIS	Sté Sucrière et Agricole de Vauciennes	IRIS Villeneuve d'Ascq	1978

	Address	Map	Scale (Size)	System	Constructor	Monitoring	Start-up
471	Zuccherificio SIIZ I-ARGELATO (BO)	C3	Pilot scale (150 m^3)	Anaerobic filter	AGIPGIZA Reggio Emilia	ASSORENI Rome	1980
472	Zuccherificio SIIZ I-ARGELATO (BO)	C3	Full scale (2700 m^3)	Anaerobic filter	AGIPGIZA Reggio Emilia	ASSORENI Rome	1982
473	CSM beet sugar factory Markkade, 29 NL-BREDA	C6	Full scale (1300 m^3)	Upflow without carrier (32 °C)	SPAANS BV HOOFDDORP	CSM Biothane Division Amsterdam	10.1980
474	CSM beet sugar factory Markkade, 29 NL-BREDA	C6	Pilot scale (30 m^3)	Upflow without carrier	CSM Amsterdam	CSM Amsterdam	09.1976
475	R. A. de Jong NL-FRANEKER	C2	Full scale (100 m^3)	Upflow without carrier	SPAANS BV HOOFDDORP	CSM Biothane Division Amsterdam	1981
476	Suiker Unie NL-GRONINGEN	D2	Full scale (1450 m^3)	Upflow without carrier	ESMIL HUBERT BV, Sneek	ESMIL HUBERT BV, Sneek	10.1979
477	CSM beet sugar factory NL-HALFWEG	B4	Full scale (800 m^3)	Upflow without carrier	CSM/SPAANS BV HOOFDDORP	CSM Biothane Division Amsterdam	10.1978
478	Suiker Unie NL-PUTTERSHOEK	B5	Full scale (1700 m^3)	Upflow without carrier	SPAANS BV HOOFDDORP	CSM Biothane Division Amsterdam	1981
479	Suiker Unie NL-ROOSENDAAL	B6	Full scale (1000 m^3)	Upflow without carrier	SPAANS BV HOOFDDORP	Owner	1983
480	CSM beet sugar factory NL-VIERVERLATEN	D2	Full scale (1800 m^3)	Upflow without carrier	SPAANS BV HOOFDDORP	CSM Biothane Div., Amsterdam	1981

	Address	Map	Scale (Size)	System	Constructor	Monitoring	Start-up
481	British Sugar Corp. UK-BURY ST EDMUNDS, Suffolk	G78	Full scale	Anaerobic contact	SORIGONA- Staffanstorp (Sweden)		10.1982
		IV.2. SUGARY WASTES (BEET PULPS-BEET TRASHES)					
482	Raperie de Marquillies F-59274 MARQUILLIES	E1	Full scale (2 x 33 m^3)	Anaerobic contact	Ets J. HAZE	IRIS + INRA Villeneuve d'Ascq	10.1980
		IV.3. SUGARY + DISTILLERY WASTEWATERS					
483	Sucrerie de Goussainville F-95190 GOUSSAINVILLE (Val d'Oise)	E2	Full scale (830 m^3)	IRIS (35-40 °C)	Sté BEGHIN SAY	IRIS Villeneuve d'Ascq	1979
		IV.4. DISTILLERY WASTEWATERS					
484	Gist en Spiritusfabriek Bruggeman, N.V. Langerbruggekai, 37 B-9000 GENT		Pilot scale (8 m^3)	Upflow without carrier Two-step (30-35 °C)	DE KONINCKX/ VANDEZANDE Wilrijk	De Koninckx/ RUG + Bruggeman Gent	10.1980
485	Fa Kraul u. Wilkening u. Stelling KG, Lohweg D-3000 HANNOVER- ANDERTEN	C4	Full scale (2500 m^3)	Anaerobic contact	Inst. für Siedlungs- wasserwirt- schaft, Univ. Hannover (TH) + Aqua consult	None	1977 (use of gas)
486	Sté Interprofessionnelle de l'Armagnac F-32100 CONDOM (Gers)	D7	Pilot scale (20 m^3)	Anaerobic filter (35 °C)	S.G.N.	INRA Narbonne	02.1982

	Address	Map	Scale (Size)	System	Constructor	Monitoring	Start-up
487	Distillerie coopérative agricole de la région d'Auvernaux F-91830 LE COUDRAY MONTCEAUX	E3	Full scale (1500 m^3)	IRIS (35 °C)	ERPAC	IRIS Villeneuve d'Ascq	1981
488	Distillerie de Lieusaint F-77127 LIEUSAINT	E3	Full scale (2 x 2000 m^3)	IRIS	Sté BEGHIN SAY	IRIS Villeneuve d'Ascq	1980
489	Erkis Industry GR-MEGARA		Full scale (500 m^3)	Anaerobic contact (50 °C)	Laboratory of Organic Chemistry Technology, Technical Univ. of Athens	Lab. of Organic Chemistry Technology, Technical Univ. of Athens	01.1983
490	Distilleria Bonollo I-ANAGNI (FR)	D5	Full scale (6000 + 2500 m^3)	Completely mixed	RPA Fontana	RPA Fontana	1982
491	Distilleria GIO-BUTON I-CASAPULLA (CE)	E5	Full scale (300 m^3)	Completely mixed	RPA Fontana	RPA Fontana	04.1981
492	Distilleria Neri Gaspare I-FAENZA (RA)	D3	Full scale (2 x 6000 m^3)	Completely mixed	RPA Fontana	RPA Fontana	04.1981
493	Distilleria Gallo I-FERRARA (FE)	C3	Full scale (3000 m^3)	Completely mixed	RPA Fontana	RPA Fontana	1982
494	Distilleria Italvinum I-FERRARA (FE)	C3	Full scale (4000 m^3)	Completely mixed	RPA Fontana	RPA Fontana	05.1982

	Address	Map	Scale (Size)	System	Constructor	Monitoring	Start-up
495	Distilleria Frontenac I-MARTIN SICURO (TE)	D4	Full scale (300 m^3)	Completely mixed	RPA Fontana	RPA Fontana	05.1981
496	Distilleria Di Lorenzo I-PONTE VALLECEPPI (PG)	D4	Full scale (2500 m^3)	Completely mixed	RPA Fontana	RPA Fontana	12.1981
497	Distilleria Toschi I-VIGNOLA (MO)	C3	Full scale (3000 m^3)	Completely mixed	RPA Fontana	RPA Fontana	01.1982
498	Distilleria M.G. I-VOLTANA (RA)	C3	Full scale (2000 m^3)	Completely mixed	RPA Fontana	RPA Fontana	10.1975
499	Distilleria Del Bosco I-VOLTANA (RA)	C3	Full scale (1000 m^3)	Completely mixed	RPA Fontana	RPA Fontana	05.1975
500	Ceimici Teoranta Dundalk IRL-Co-LOUTH	C6	Pilot scale (14 m^3)	Anaerobic filter (35 °C)	BIOMASS Ltd. Galway	BIOMASS Ltd. Galway	03.1982
501	Zuid-Ned. Gist en Spiritus Fabriek NL-BERGEN OP ZOOM	B6	Full scale (750 m^3)	Upflow without carrier	SPAANS BV Hoofddorp	Owner/CSM Biothane Div. Amsterdam	1981
502	Long John International 1780 London Rd UK-GLASGOW		Pilot scale (10 m^3)	Anaerobic contact	BIOMECHANICS Ltd. Ashford		

IV.5. STARCH PRODUCTION WASTEWATERS

	Address	Map	Scale (Size)	System	Constructor	Monitoring	Start-up
503	Usine TENSTAR 36, Quai de Brazza F-33015 BORDEAUX Cédex	C6	Full scale (2 x 2000 m^3)	Anaerobic contact Continuous (37 °C)	BIOMECHANICS Ltd., United Kingdom	None	1976

	Address	Map	Scale (Size)	System	Constructor	Monitoring	Start-up
504	Wheat Industries Ltd. Ringaskiddy IRL-Co-CORK	B8	Full scale (2200 m^3)	Upflow without carrier	PAQUES The Netherlands	Wheat Ind. Co-Cork	1983
505	Smiths Chips NL-BROEK OP LANGEDIJK	B3	Full scale (420 m^3)	Upflow without carrier	SPAANS BV HOOFDDORP	CSM Biothane Division Amsterdam	1983
506	Zetmeelbedrijven (ZBB) NL-KOOG AAN DE ZAAN	B4	Full scale (800 m^3)	Upflow without carrier	HEIDEMIJ/ LEDOUX Arnhem	Heidemij Arnhem	1980
507	Latenstein Wessanen NL-NIJMEGEN	D5	Full scale (640 m^3)	Upflow without carrier	Spaans BV Hoofddorp	CSM Biothane Division Amsterdam	1983
508	TENSTAR PRODUCTS Ltd. Hythe Road UK-ASHFORD (Kent)	G9	Full scale (3000 m^3)	Completely mixed continuous (35-37 °C)	BIOMECHANICS/ TENSTAR PRO-DUCTS Ltd. Ashford	None	1976
			III.6. <u>PECTIN INDUSTRY WASTEWATERS</u>				
509	Fa Pomosin GmbH v. Herwarth Str. D-2443 GROSSENBRODE	D1	Full scale (3000 m^3)	Anaerobic contact	AQUA CONSULT Hannover	Inst. für Siedlungs-wasserserwirt-schaft Univ. Hannover	08.1982 (use of gas)
510	Københavns Pektinfabrik A/S, Ved Banen, 12 DK-4623 LILLE SKENSVELD	F5	Full scale (2145 m^3 + 1473 m^3)	Anaerobic contact	MØLLER & JOCHUMSEN A/S	Water Quality Inst. of Hør-sholm (routine)	1979

	Address	Map	Scale (Size)	System	Constructor	Monitoring	Start-up
511	Usine UNIPECTINE 24, rue de la Jicquelaie F-35600 REDON	C4	Full scale (3000 m^3)	Anaerobic contact (37-38 °C)	KRÜGER Copenhagen	UNIPECTINE Redon	1979
			IV.7. CONFECTIONERY WASTEWATERS				
512	Swizzels Matlow Confec- tioners, New Mills UK-MANCHESTER	E7	Pilot scale (10 m^3)	Plug-flow - Anaerobic filter (35-38 °C)	MASS TRANSFER Ltd.	Pollution Research Unit of UMIST, Manchester	01.1982
			IV.8. SLAUGHTER-HOUSE WASTEWATERS				
513	Slagteri-Region Syd DK-6372 BYLDERUP BOV	B6	Pilot scale (10 m^3)	Anaerobic filter	Danish Fermen- tation Ind. Glostrup		1981
514	Slagteri-Region Syd DK-6372 BYLDERUP BOV	B6	Full scale (400 m^3)	Anaerobic filter	Danish Fermen- tation Ind. Glostrup		1983
515	Residentie Slachthuis BV NL-DEN HAAG	B5	Full scale (800 m^3)	Upflow without carrier	TAUW Deventer	TAUW Deventer	1983
			IV.9. SLAUGHTER-HOUSE WASTES				
516	Abattoir de Charleroi Place de l'Abattoir B-6000 CHARLEROI		Pilot scale (1 m^3)	Completely mixed Semi-continuous (35 °C)	Local tradesmen	GEBI-UCL Louvain- la-Neuve	01.1980 to 07.1981
517	Abattoir de Ciney B-5300 CINEY		Pilot scale (10 5 m^3)	Completely mixed Semi-continuous (35 °C)	Owner	AIDR + GEBI UCL, Louvain- la-Neuve	01.1980 to 06.1981

	Address	Map	Scale (Size)	System	Constructor	Monitoring	Start-up
518	Abattoir municipal de Castres F-81100 CASTRES	E7	Full scale (45 m^3)	Batch (35 $\pm$ 3 °C)	ARES + local tradesmen Castres	ARES Castres	12.1980
519	Abattoir de St Lô Normandie Viande F-SAINT-LO		Pilot scale (4 m^3)		Sté BERTIN Plaisir	Sté BERTIN Plaisir	1982
			IV.10. MILK PROCESSING WASTEWATERS				
520	Carberry Milk Products Ltd. Ballineen, IRL-CO-CORK	B8	Pilot scale	Anaerobic filter	BIOMASS Ltd. Galway	BIOMASS Ltd. Galway	1982
521	Golden Vale Creamery Charleville IRL-CO-CORK	B8	Pilot scale	Anaerobic filter	HANLEY Brothers	Microbiology Dept. U.C.G. Galway	01.1978 to 01.1979
			IV.11. POTATO PROCESSING WASTEWATERS				
522	SANALCO Skovlund DK-6823 ANSAGER	A5	Pilot scale (10 m^3)	Anaerobic filter	Danish Fermen- tation Ind. Glostrup		1981
523	SANALCO Skovlund DK-6823 ANSAGER	A5	Full scale (400 m^3)	Anaerobic filter	Danish Fermen- tation Ind. Glostrup	1983	
524	Mac Cain, Alimentaire SARL Z.I. d'Harnes F-62440 HARNES	E1	Full scale (26000 m^3)	Accumulation with recirculation of the mixed liquor	ADI (Canada) + CARONI	ADI Ltd.	08.1981

	Address	Map	Scale (Size)	System	Constructor	Monitoring	Start-up
525	Frites d'Or NL-BERGEN OP ZOOM	B6	Full scale (1400 m^3)	Upflow without carrier	PAQUES BV Balk	Vereniging Krachtwerk-tuigen Amersfoort	1982
526	AVeBe NL-DE KRIM	D3	Full scale (5000 m^3)	Upflow without carrier	SPAANS BV HOOFDDORP	Owner/CSM Biothane Div. Amsterdam	1981
527	AVeBE NL-FOXHOL	D2	Full scale (1100 m^3)	Upflow without carrier	SPAANS BV HOOFDDORP	Owner	1980
528	Frites specialist NL-LOMM	D5	Full scale (400 m^3)	Upflow without carrier	ESMIL HUBERT Sneek	ESMIL HUBERT BV/CSM	09.1980
529	Farm frites NL-OUDENHOORN	B5	Full scale (240 + 700 m^3)	Upflow without carrier	ESMIL HUBERT Sneek	ESMIL HUBERT BV Sneek	1980
530	AVIKO NL-STEENDEREN	D4	Full scale (1500 m^3)	Upflow without carrier	ESMIL HUBERT Sneek	ESMIL HUBERT Sneek	12.1981
531	RIXONA NL-WARFFUM	D1	Full scale (4 x 50 m^3)	Upflow without carrier	Owner	Owner	1977
532	KUIBO NL-WASPIK	C5	Full scale (300 m^3)	Upflow without carrier	PAQUES BV Balk	Ver. Kracht-werktuigen Amersfoort	1981
			IV.12. <u>FLAX INDUSTRY WASTEWATERS</u>				
533	Flax industry M. Kinds B-9800 GOTTEM	B2	Full scale (12 m^3)	Upflow without carrier, one-step (35 °C)	DE KONINCKX VANDEZANDE Wilrijk	De Koninckx + RUG Gent	09.1980

	Address	Map	Scale (Size)	System	Constructor	Monitoring	Start-up
			IV.13. VEGETABLE CANNING INDUSTRY WASTEWATERS				
534	Ets Bonduelle Usine de Renescure F-59173 RENESCURE	E1	Full scale (2 x 2500 m^3)	Anaerobic contact Continuous (35 ± 5 °C)	DEGREMONT + BONDUELLE Rueil- Malmaison	INRA Villeneuve d'Ascq	07.1981 and 04.1982
			IV.14. OLIVE OIL WASTEWATERS				
535	Research Institute of IRSA-CNR I-70123 BARI		Pilot scale (8 + 2 x 5 m^3)			Inst. of IRSA- CNR Bari	
			IV.15. ENZYME INDUSTRY WASTEWATER				
536	Usine CAPPA Rue de Liège F-59121 PROUVY (Nord)	F1	Pilot scale (20 m^3)	Upflow without carrier (36 °C)	CAPPA Prouvy	INRA Villeneuve d'Ascq	02.1982
			IV.16. BREWERY WASTEWATERS				
537	Palm Steenhuffel Brouwerij nv Steenhuffeldorp, 3 B-2900 LONDERZEEL	C2	Full scale	Upflow without carrier	O. DE KONINCKX S.A. Wilrijk		1983
538	Sébastien Artois Avenue Roger Salengro F-59280 ARMENTIERES	E1	Full scale	Two-step	DEGREMONT Rueil Malmaison		1983
539	Union des Brasseries F-13000 MARSEILLE	G7	Pilot scale (4-5 m^3)	Anaerobic contact	Sté BERTIN Plaisir	Sté BERTIN Plaisir	1982-83

	Address	Map	Scale (Size)	System	Constructor	Monitoring	Start-up
			IV.17. CITRIC ACID FERMENTATION WASTEWATERS				
540	J. & E. Sturge Ltd. Denison Rd, UK-SELBY, N. Yorks	F6	Full scale ($5100\ m^3$)		Ames Crosta Heywood		1970
			IV.18. LEATHER FACTORY WASTEWATERS				
541	Lederfabrik Helen D-3452 BODENWERDER		Full scale		Steinmann u. Ittig GmbH u Co Kg Ingenieur u. Behälterbau Minden	None	1983 (use of gas)
			IV.19. WOODEN PLATES FACTORY WASTEWATERS				
542	Herzberger Plattenwerke D-342 HERZBERG	D4	Full scale ($2 \times 4000\ m^3$)	Anaerobic contact	Steinmann u. Ittig GmbH u Co Kg Ingenieur - u. Behälterbau Minden	Niedersächsisches Wasseruntersuchungsamt Hildesheim	06.1982 (use of gas)
			IV.20. PICKLES CANNING WASTEWATERS				
543	Fa. Hengstenberg Werk Fritzlar D-358 FRITZLAR	C5	Full scale ($2 \times 2500\ m^3$)	Anaerobic contact	Owner	None	1977 (use of gas)
			IV.21. PAPER INDUSTRY WASTEWATERS				
544	ACI NL-DEN HAAG		Full scale ($800\ m^3$)	Upflow without support			1983

	Address	Map	Scale (Size)	System	Constructor	Monitoring	Start-up
			IV.22. FISH - CANNERY WASTEWATERS				
545	Van der Plasse NL-YERSEKE	A6	Full scale (100 m^3)	Upflow without carrier	TAUW Deventer	TAUW Deventer	1981
			IV.23. ONION PROCESSING WASTEWATERS				
546	DGI DK-2600 GLOSTRUP	F5	Pilot scale (10 m^3)	Anaerobic filter	Danish Fermentation Ind. Glostrup	Danish Fermentation Ind. Glostrup	1982-83

APPENDIX IV. ADDRESSES OF CONSTRUCTORS, CONSULTANTS, LABORATORIES AND ORGANIZATIONS INVOLVED IN BIOMETHANATION

Addresses are listed by country and then alphabetically under each heading.

1° BELGIUM

Constructors & consultants :

- ACEC
  (Ateliers de Construction Electrique de Charleroi)
  B.P. 4
  B-6000 CHARLEROI

- AIDR
  (Association Internationale de Développement Rural)
  Rue de la Science, 9
  B-1040 BRUXELLES

- Biométhane, Sté Coop.
  Rue des Beguinettes, 9
  B-1170 BRUXELLES

- Bio-watt pvba
  Brusselsesteenweg, 363
  B-1881 MERCHTEM-BRUSSEGEM

- EKW pvba
  Kiefhoekstraat, 1A
  B-3588 EKSEL

- IVPO Delegatie Vlaanderen
  (Intenationale Vereniging voor Plattelandsontwikkeling)
  Marcus van Vaernewijckstraat, 25
  B-9910 MARIAKERKE-GENT

- Studiebureau O. De Koninckx
  Valkenveld, 1-12
  B-2610 WILRIJK

Laboratories & organizations :

- Cebedeau (Edeline)
  Rue A. Stevart, 2
  B-4000 LIEGE

- Centre Agronomique de Recherches Appliquées du Hainaut (CARAH) (Dessart)
  Institut Agricole du Hainaut
  Rue Paul Pastur, 11
  B-7800 ATH

- Faculté de Médecine, Département de Microbiologie (Joassin)
  Université de Liège (ULg)
  Sart Tilman
  B-4000 LIEGE

- Faculté des Sciences Economiques et Sociales (Schepens)
  Facultés Universitaires Notre-Dame de la Paix (FUNDP)
  Rempart de la Vierge, 8
  B-5000 NAMUR

- Laboratorium voor Algemene en Toegepaste Microbiële Ecologie (Verstraete)
  Rijksuniversiteit - Gent (RUG)
  Coupure Links 653
  B-9000 GENT

- Laboratorium voor Analytische Scheikunde (Elskens)
  Vrije Universiteit Brussel, (VUB)
  Pleinlaan, 2
  B-1050 BRUSSEL

- Laboratoire d'Automatique et d'Analyse des Systèmes (Installé)
  Université Catholique de Louvain, (UCL)
  Place du Levant, 3 (Bâtiment Maxwell)
  B-1348 LOUVAIN-LA-NEUVE

- Unité de Génie Biologique (Nyns & Naveau)
  Université Catholique de Louvain (UCL)
  Place Croix du Sud, 1/9
  B-1348 LOUVAIN-LA-NEUVE

2° <u>DENMARK</u>

<u>Constructors & consultants</u> :

- Agri-Nova A/S (3)
  Marskvej, 29
  DK-4700 NAESTVED

- Agro-Metan (1)
  Brogade, 1, Seppersund
  DK-9240 NIBE

- Andersen & Fog Ltd., (1)
  Tranekaerparken, 17-19
  DK-8240 RISSKOV

---

(1) Turn-key contractors
(3) Consultants

- August Olsens Eftf. I/S (2) (flexible membrane)
  Strandlodsvej, 20
  DK-2300 COPENHAGEN S

- Biodangas (3)
  Stensbjergvej, 3, Lind
  DK-7400 HERNING

- Carl Bro A/S (3)
  Granskoven, 8
  DK-2600 GLOSTRUP

- CowiConsult A/S (3)
  Teknikerbyen, 45
  DK-2830 VIRUM

- Crone & Koch A/S (3)
  Faelledvej, 1
  DK-8800 VIBORG

- Dansk Electromotorlager (2) (generator sets)
  Kastanie Alle, 8
  DK-2720 VANLØSE

- Dansk Gaerings Industri (3)
  Fabriksparken, 58
  DK-2600 GLOSTRUP

- A/S Dansk Wimex (2) (gas engine)
  Køgevej, 32
  DK-2630 TASTRUP

- Højbogard Biogasanlaeg A/S (1)
  Industrivej, 9
  DK-5580 NØRRE ABY

- Hørlych & Kongsted (1)
  DK-6580 VAMDRUP

- Jysk Biogas (1)
  DK-9640 FARSØ

- I. Krüger A/S (1) (2) (3)
  Rymarksvej, 108
  DK-2100 COPENHAGEN Ø

- Manheim Diesel A/S (2) (gas engine)
  Speditørvej, 5
  DK-2450 COPENHAGEN SV

- Motorfabrikken Bukh (2) (gas engine)
  DK-4400 KALUNDBORG

---

(1) Turn-key contractors
(2) Suppliers of components
(3) Consultants

- Nordisk Gummibadsfabrik (2) (flexible membrane)
  Sedding Ringvej
  DK-6700 ESBJERG

- Roheikco (2) (gas engine)
  DK-5560 AARUP

- Samfundsteknik A/S (3)
  Pjentedamsgade, 21
  DK-5000 ODENSE C

- Saxonia (2) (TOTEM)
  Borgmester Christiansensgade, 58
  DK-2450 COPENHAGEN SV

Laboratories & associations

- Aalborg University Centre (Knud Dahl)
  Postbox, 159
  DK-9100 AALBORG

- Crone & Koch, Consulting Engineers (H.C. Willumsen)
  Faelledvej, 1
  DK-8800 VIBORG

- Danish Biogastechnical Association (Poul Keller)
  SJF Bygholm
  DK-8700 HORSENS

- Dansk Biogasteknisk Forening
  (Danish Association for Biogas Technology)

  Jens Jensen (Chairman)
  Horsbølvej, 1, Rejsby
  DK-6780 SKAERBAEK

  Susanne Hansen (secretary)
  The Technological Institute
  Department of Chemistry
  Gregersensvej
  DK-2630 TASTRUP

- D.E.L. Manufacturers of Electrical Equipment (T. Kongstad)
  Kastanie Alle, 8
  DK-2720 VANLØSE

- Højbogard Biogas Plant Nørre Aby A/S (Niels Thaegaard)
  Industrivej, 9
  DK-5580 NØRRE ABY

- Institute of Technical Biochemistry (Mogens Hølger Jørgensen)
  Technical University of Denmark
  Building 223
  DK-2800 LINGBY

---

(2) Suppliers of components
(3) Consultants

- Laboratory of Energetics (Erik Kofoed)
  Technical University of Denmark
  Building 403
  DK-2800 LINGBY

- Laboratory of Sanitary Engineering (Mogens Henze)
  Technical University of Denmark
  Building 115
  DK-2800 LINGBY

- Project Godrum
  Viborgvej, 188
  DK-8210 AARHUS V

- Revninge Co-operative Energy Society
  Revninge, Kertminde Municipality
  c/o R. Pedersen
  Pilevaenget, 5
  DK-5300 KERTMINDE

- Royal Veterinary and Agricultural University (Josef Zubr)
  Thorvaldsensvej, 40
  DK-1871 KØBENHAVN V

- STUB c/o Carl Bro A/S (Gerhardt Grøn)
  Granskoven, 8
  DK-2600 GLOSTRUP

3° FRANCE

Constructors & consultants :
(non-exhaustive listing, established from AFME papers (Agence Française pour la Maîtrise de l'Energie))

- Air Liquide
  Division Biogas
  Avenue Carnot, 57
  F-94500 CHAMPIGNY SUR MARNE

- ARES
  Application de Recherches sur l'Energie et la Société
  Rue Louise Beaupuy, 12
  F-81100 CASTRES

- Bendel
  Rue de Nancy
  F-57430 SARRALBE

- Bertin
  Allée Gabriel Voisin
  B.P. 3
  F-78370 PLAISIR

- Bioflor
  Avenue du XXème Corps, 45
  F-54000 NANCY

- Cossoul Agrinergie France
  Rue Marie Maurel, 1
  F-15001 AURILLAC

- Degrémont
  Avenue du 18 juin 1940, 183
  F-92508 RUEIL MALMAISON Cedex

- EMC
  Entreprises Minières et Chimiques
  PEC-Engineering
  Rue Jeanne d'Arc, 62
  F-75646 PARIS Cedex 13

- Erpac
  Rue Carpeaux, 85
  F-59290 WASQUEHAL

- Induferma
  Rue de Bercy, 185
  F-75012 PARIS

- Inovena
  Rame
  F-09230 SAINTE-CROIX-VOLVESTRE

- La Caussenarde
  Lavernhe
  F-12150 SEVERAC LE CHATEAU

- Marumo
  Rue de la Forêt
  Hartmanns Willer
  F-68500 GUEBWILLER

- Mécanique des Landes
  Zone Industrielle de Tarnos
  F-40220 TARNOS

- Methaneco
  Myans
  F-73800 MONTMELIAN

- Mida-Tra
  Castelnau Barbarens
  F-32450 SARAMON

- Montavon
  Ets Jean Montavon
  Z.I. Faubourg de Belfort
  F-90100 DELLE

- OTV
Omnium de Traitement et de Valorisation
"Le Doublon"
Avenue Dubonnet, 11
F-92407 COURBEVOIE Cedex

- SAPEC
Socété Aquitaine pour l'Etude et le Contrôle des Eaux
Avenue A. Nobel
Zone Industrielle Indusnor
F-64000 PAU

- SCOPIC
Rue du Quai, 12
F-34220 SAINT-PONS

- SCTR
Société de Chaudronnerie et Tuyauterie Rhodanienne
En Milon Brindas
B.P. 2
F-69290 CRAPONNE

- SEBM
Société d'Exploitation de Brevets et de Matériels
Avenue Pasteur, 73
F-59130 LAMBERSART

- SEITHA
Société d'Etudes et d'Installations Thermique et Aéraulique
Rue du Commandant Lherminier, 3
B.P. 5046
F-69601 VILLEURBANNE Cedex

- SETAL
Société d'Etudes Techniques Aquitaine Languedoc
Avenue des Lilas, 130
F-64000 PAU

- Serru
Ets Maurice Serru SA
Route de sablé
F-53200 CHATEAU GONTIER

- Sevia Batigral
Energie Nouvelles
Chemin de la Reverbière
Fitilieu
F-38490 LES ABRETS

- SGN
Saint Gobain Techniques Nouvelles
Rue des Hérons, 1
Montigny le Bretonneaux, B.P. 30
F-78184 SAINT QUENTIN EN YVELINES Cedex

- Sigoure Frères
  Zone Industrielle Riorges
  B.P. 40
  F-42309 ROANNE Cedex

- SOVAP-SOMAP
  Rue Ampère, 30
  F-85000 LA ROCHE SUR YON

- SSCM
  Sté Stéphanoise de Constructions Mécaniques
  Rue Scheurer Kestner, 10, B.P. 519
  F-42007 SAINT ETIENNE Cedex

- Zodiac
  Quai du Président Carnot, 61
  F-92210 SAINT CLOUD

Laboratories & associations :

1) Involved in the monitoring of biogas plants within the frame of the joint monitoring campaign of AFME (Agence Française pour la Maîtrise de l'Energie) and GIDA (Groupement Interinstituts pour l'Etude et l'Utilisation des Déjections Animales).

- REGION AQUITAINE
  Chambre d'agriculture
  Place de la Mairie, 1
  F-47000 AGEN

- DEPARTEMENTS AVEYRON-CANTAL
  Institut Technique du Porc
  Station expérimentale - Les Cabrières
  F-12200 VILLEFRANCHE-DE-ROUERGUE

- DEPARTEMENT MAINE ET LOIRE
  INRA Station Agronomique de Loire Atlantique
  Boulevard Victor Hugo, 20
  F-44200 NANTES

- REGION ILE-DE-FRANCE
  GIDA Maison Nationale des Eleveurs
  Rue de Bercy, 149
  F-75012 PARIS

- REGION LANGUEDOC
  INRA Service d'Etudes des Sols
  Centre de Recherches de Montpellier
  Place Viala, 9
  F-34060 MONTPELLIER

- REGION RHÔNE-ALPES
  BCMEA Chambre d'Agriculture du Rhône
  Place Jensoul, 4
  F-69002 LYON

	Address	Map	Scale (Size)	System	Constructor	Monitoring	Start-up
080	L. Majoie Chaussée de Charleroi B-5921 THOREMBAIS-LES-BEGUINES	D3	Full scale (136 m^3)	Completely mixed Int. loaded (35 °C)	Local tradesmen	GEBI-UCL on behalf COMET Louvain-la-Neuve	02.1982
081	H. Schulze Mallinckrodtweg, 25 D-4730 AHLEN	B4	Full scale (50 m^3)	Completely mixed (35 °C)	AFAG Basel (CH)	None	12.1979
082	M. Haindl, Ederhof Arnstorfer str. 4 D-8353 ALTENMARKT	F8	Full scale (250 m^3)	Mixed (hydraulic) (33 °C)	BVT Bregenz (A)	Landtechnik Weihenstephan	12.1981
083	V. Reden-Lütcken Georgenau D-7263 BAD LIEBEN-ZELL-MOTTLINGEN	C8	Full scale (80 m^3)	Batch	Owner	Universität Hohenheim	1979
084	K. Abbt Sebastianstrasse, 1 D-8091 BATZENHOFEN (bei Augsburg)	D8	Full scale (42 m^3)	Completely mixed (35 °C)	Owner	None	05.1980
085	H. Schillingmann D-4576 BERGE DALVERS	B3	Full scale (68 m^3)	Plug-flow (15-30 °C)	Owner	FAL Braunschweig	03.1982
086	J. Kohnle D-7091 BERGHEIM (Tannhausen)	D7	Full scale (62 m^3)	Plug-flow + mixing (38 °C)	LIPP GmbH Tannhausen	None	11.1980
087	Bundesforschungsanstalt für Landwirtschaft (FAL) Bundesallee, 50 D-3300 BRAUNSCHWEIG	D4	Full scale (100 m^3)	Partially mixed (30-55 °C) Experimental	Owner	Inst. für Technologie Braunschweig	07.1980

- INRA Station de Technologie Alimentaire
  CERTIA rue J. Guesde, 369
  F-59651 VILLENEUVE D'ASCQ Cédex

- INRA
  Station d'Oenologie et de Technologie Végétale
  Avenue du Général de Gaulle
  F-11104 NARBONNE Cedex

- INSA
  Institut National des Sciences Appliquées
  Avenue de Ranguenil
  F-31077 TOULOUSE Cedex

- INA
  Institut National Agronomique
  Thiverval
  F-78850 GRIGNON

- IRAT
  Institut de Recherches Agronomiques Tropicales
  Montferrier sur Lez, B.P. 5035
  F-34032 MONTPELLIER Cedex

- IRChA
  Institut National de Recherche Chimique Appliquée
  B.P. 1
  F-91710 VERT-LE-PETIT

- IRIS
  Institut de Recherche de l'Industrie Sucrière
  CERTIA, Rue J. Guesde, 369
  F-59650 VILLENEUVE D'ASCQ

- IFCC
  Institut Français du Café, du Cacao et autres plantes stimulantes
  Montferrier sur Lez, B.P. 5035
  F-34032 MONTPELLIER Cedex

- INPL
  Institut National Polytechnique de Lorraine
  B.P. 3308
  F-54014 NANCY Cedex

- IRH
  Institut de Recherches Hydrologiques
  Rue Ernest Bichat, 10
  F-54000 NANCY

- ITAVI
  Institut Technique de l'Aviculture
  Rue du Rocher, 12
  F-75008 PARIS

- ITCF
  Institut Technique des Céréales et des Fourrages
  Boigneville
  F-91710 MAISSE

- ITEB
  Institut Technique de l'Elevage Bovin
  rue de Bercy, 149
  F-75579 PARIS Cédex 12

- ITG
  Institut Technique du Gruyère
  F-70190 RIOZ

- ITP
  Institut Technique du Porc
  Bd de la gare, 34
  F-31079 TOULOUSE Cedex

- Université de Compiègne
  B.P. 233
  F-60206 COMPIEGNE

3) Not involved in monitoring of biogas plants but working on or interested in biomethanation

- ANERA
  Association Normande d'Economie Rurale Appliquée
  Place Vernoise
  F-14000 CAEN

- ANRED
  Association Nationale pour la Récupération et l'Elimination des Déchets
  Square Lafayette, B.P. 406
  F-49004 ANGERS Cédex

- CNRS
  Centre National de la Recherche Scientifique
  Laboratoire de Chimie Bactérienne
  Chemin Joseph Aiguier, 31
  13009 MARSEILLE

- Coopérative des Agriculteurs du Morbihan
  F-56000 VANNES

- GDF
  Gaz de France
  Direction des Etudes et Techniques Nouvelles
  rue d'Alsace
  F-92300 LEVALLOIS

- ORSTOM
  Office de la Recherche Scientifique et Technique d'Outre-Mer
  Laboratoire de Microbiologie
  IRChA - B.P. 1
  F-91710 VERT-LE-PETIT

- Sté Guyomarc'h
  B.P. 235
  F-56000 VANNES

- TIMAC
  B.P. 158
  F-35400 SAINT MALO

4° FEDERAL REPUBLIC OF GERMANY

Constructors & consultants

a) For agricultural biogas plants

- AFAG
  Agro-Fermenttechnik AG
  Homburger Strasse, 47
  CH-4052 BASEL/SCHWEIZ
  und
  Mallinckrodtweg, 25
  Haus Küchen
  D-4730 AHLEN

- Barth, B.
  Agrartechnik
  D-7527 KRAICHTAL-UNTERÖWISHEIM

- BVT (Biologische Verfahrenstechnik GmbH)
  Bergstrasse, 8
  D-6900 BREGENZ/ÖSTERREICH

- Chemotechnik GmbH Kunststoff und Apparatebau KG
  Postfach 169
  D-8999 WEILER/ALLGAU

- FIMA
  Fischachtaler Maschinenbau GmbH + Co
  D-7164 OBERSONTHEIM

- Gebken-Biogasanlagen GmbH + Co KG
  Hatterlandstrasse, 44
  D-2904 HATTEN-TWEELBÄKE

- Geiger, Karlheinz
  Römerstrasse, 11
  D-8000 MÜNCHEN 80

- Gesellschaft für Alternative Energieerzeugung mbH
  Hof am Storchennest
  D-8711 MICHELFELD

- GFIT (Gesellschaft für Interdisziplinare Technologie mbH)
  Pfaffenhofener Strasse, 2
  D-7914 ROTH-PFAFFENHOFEN

- Henze-Harvestore GmbH
  Nordbrögger Strasse, 9-11
  D-4700 HAMM 3 (Lerche)

- Joba-Biogas
  Postfach
  D-2830 BASSUM

- Landwirtschaftlicher Anlagenbau Wilfried Schraufstetter
  Schlossstrasse, 5
  D-80045 ISMANING

- Lipp GmbH
  D-7091 TANNHAUSEN/OSTABLBKREIS

- Metallbau Schröder
  D-2300 RENDSBURG

- Müller, Werner
  D-8200 ROSENHEIM/INN

- Ökoplan
  Eisenbüttler Strasse, 13
  D-3300 BRAUNSCHWEIG

- Ökotherm Vertriebs-GmbH
  Hauptstrasse, 1
  D-2351 HAGEN

- Revat
  Ingenieurbüro für Regenerative Verfahrenstechnik
  Am Schulberg, 6a
  D-2114 DIERSTORF

- Steinmann + Ittig
  Ingenieur- und Behältherbau GmbH + CO KG
  Postfach 3110
  Minderheideweg, 2
  D-4950 MINDEN

- UTEC
  Ingenieurbüro für Entwicklung und Anwendung Umweltfreundlicher Energietechnik
  Tollendorf, 9
  D-3139 GÖHRDE 1

- Württemberg Landeszentr. Raiffeisen-Genossenschaft WLZ
  Undisweiler
  D-7968 SAULGAU 1

b) <u>For industrial biogas plants</u>

- AB-Sorigona c/o AC. Biotechnics (Industrial Biotechnology)
  Box 4
  S-23200 ARLÖV (SWEDEN)

- Braunschweigische Maschinenbauanstalt (BMA)
  Am Alten Bahnhof, 5
  D-3300 BRAUNSCHWEIG

- Institut für Siedlungswasserwirtschaft, Universität Hannover (TH)
  Welfengarten, 1
  D-3000 HANNOVER in cooperation with :

  Aqua Consult
  Lange Laube, 29
  D-3000 HANNOVER

- Steinmann u. Ittig GmbH u. Co. KG
  Ingenieur-u. Behälterbau
  Minderheideweg, 2
  Postfach 3110
  D-4950 MINDEN/WESTF.

- Gebrüder Sulzer AG
  CH-8401 WINTERTHUR (SWISS)

c) <u>For landfills</u>

- Abfallbeseitigungsbetriebe Ruhrgebiet GmbH i.G.
  Rüttenscheider strasse, 66
  D-4300 ESSEN 1

- Fuhramt der Stadt Hannover
  Karl-Wiechert-Allee, 60c
  D-3000 HANNOVER

- Gesellschaft zur Beseitigung von Sonderabfällen in Rheinland/Pfalz mbH
  Postfach 013
  D-6710 FRANKENTHAL

- Interatom
  Friedrich-Ebert-Str.
  D-5060 BERGISCH-GLADBACH 1

- Müll-Ex GmbH West u. Co. KG
  Brinkmannstr., 43
  D-2251 AHRENSHÖFT

- Büro für Kies u. Abfall W. Ryser
  Aarbord, 50
  CH-3118 UTTINGEN (SCHWEIZ)

- Tiefbauamt u. Standwerke der Stadt Pforzheim
  D-7530 PFORZHEIM

- Westerwaldkreis-Abfallbeseitigung
  Bodener Strasse
  D-5431 MOSCHHEIM

- Zander
  Wendentorwall, 19
  D-3300 BRAUNSCHWEIG

Laboratories & consultants :

Monitoring of :

a) Agricultural biogas plants

- Bayer. Landesanstalt für Tierzucht
  Abteilung, 6
  D-8011 GRUB

- Botanisches Institut der Universität München
  Menzinger Strasse, 67
  D-8000 MÜNCHEN 19

- Fachhochschule Bremerhaven Studiengang Betriebs-und Versorgungstechnik
  Columbusstrasse, 21
  D-2850 BREMERHAVEN

- Gesamthochschule Kassel
  Menzelstrasse, 13
  D-3500 KASSEL

- Institut für Landwirtschaftliche Verfahrenstechnik
  Olshausenstrasse, 40-60
  D-2300 KIEL

- Institut für Siedlungswasserwirtschaft und Abfalltechnik
  Universtät Hannover
  Welfengarten, 1
  D-3000 HANNOVER

- Institut für Technologie, Bundesforschungsanstalt für Landwirtschaft (FAL)
  Bundesallee, 50
  D-3300 BRAUNSCHWEIG

- KTBL (Kuratorium für Technik und Bauwesen in der Landwirtschaft)
  Bartningstrasse, 49
  Postfach 12 01 42
  D-6100 DARMSTADT 12

- Landtechnik Weihenstephan
  Vöttinger Strasse, 36
  D-8050 FREISING-WEIHENSTEPHAN

- Landwirtschaftsamt
  Münchberg
  D-8660 MÜNCHBERG/HOF

- Landwirtschtskammer
  Altötting
  D-8262 ALTÖTTING

- Landwirtschaftskammer Rheinland
  Westparkstrasse, 92-96
  D-4150 KREFELD

- Landwirtschaftskammer Weser-Ems
  Mars-la-Tour-Str., 1-13
  Postfach 25 49
  D-2900 OLDENBURG

- Universität Hohenheim
  Garbenstrasse, 9
  D-7000 STUTTGART 70

- Universität Oldenburg
  Ziegelhofstrasse, 82
  D-2900 OLDENBURG

b) Industrial biogas plants

- Institut f. Landw. Technologie u. Zuckerindustrie
  Langer Kamp, 5
  D-3300 BRAUNSCHWEIG

- Institut für Technologie d. Bundesforschungsanstalt für Landwirtschaft (FAL)
  Bundesallee, 50
  D-3300 BRAUNSCHWEIG

- Niedersächsische Wasseruntersuchungsamt
  Postfach
  D-3200 HILDESHEIM

c) Landfills

- Institut für Siedlungswasserbau, Wassergüte- und Abfallwirtschaft
  Bandtäle
  D-7000 STUTTGART 80

- Institut für Stadtbauwesen, Abtlg. Siedlungswasserwirtschaft,
  Universität Braunschweig
  Pockelsstrasse, 10
  D-3300 BRAUNSCHWEIG

Research on anaerobic digestion :

a) Universities :

- Inst. f. Verfahrenstechnik	TH	AACHEN
- Fachgebiet Biotechnologie	TU	BERLIN
- Inst. f. Mikrobiologie (1)	Univ.	GÖTTINGEN
- Inst. f. Siedlungswasserwirtschaft	Univ.	HANNOVER
- Inst. f. Mikrobiologie (1)	Univ.	HEIDELBERG
- Inst. f. Agrartechnik/Landesanstalt f. Landw. Maschinen- u. Bauwesen	Univ.	HOHENHEIM
- Inst. f. Tiermedizin u. Tierhygiene	Univ.	HOHENHEIM
- Biochem. Inst. d. Medizin. Fakultät (1)	Univ.	KIEL
- Inst. f. landw. Verfahrenstechnik	Univ.	KIEL
- Fakultät f. Biologie (1)	Univ.	KONSTANZ
- Fachbereich Biologie - Mikrobiologie (1)	Univ.	MARBURG
- Lehrstuhl f. Mikrobiologie (1)	Univ.	MÜNCHEN

- Botanisches Institut (1) (2) — Univ. MÜNCHEN
- Lehrstuhl f. Organ. Chemie u. Biochemie (1) — Univ. MÜNCHEN
- Lehrstuhl B f. Thermodynamik (2) — Univ. MÜNCHEN
- Inst. f. Landtechnik d. Landw. Fakultät/ Bayer. Landesanstalt f. Landtechnik — Univ. MÜNCHEN
- Biologisches Institut VII (1) — Univ. REGENSBURG

(1) participating in the interuniversity project "Methanogenic Bacteria", financed by the German Research Association (DFG)

(2) participating in the university project "Biological Chemical and Technical Fundamentals of Bioconversion" at the University of Munchen, financed by the German Research Association (DFG)

b) Research Centres

- Inst. f. Biotechnologie I (a)
  Kernforschungsanlage
  JÜLICH

- Inst. f. Biotechnologie II
  Kernforschungsanlage
  JÜLICH

- Inst. f. Bodenbiologie
  Bundesforschungsanstalt f. Landwirtschaft (FAL)
  BRAUNSCHWEIG

- Inst. f. Grenzflächen- und Bioverfahrenstechnik
  Fraunhofergesellschaft
  STUTTGART

- Inst. f. Grünland- und Futterpflanzenforschung
  Bundesforschungsanstalt f. Landwirtschaft (FAL)
  BRAUNSCHWEIG

- Inst. f. Technologie
  Bundesforschungsanstalt f. Landwirtschaft (FAL)
  BRAUNSCHWEIG

5° GREECE

Constructors & consultants

None

Laboratories & organizations :

- Agricultural College of Athens
  Lab. of Agricultural Structures
  Iera Odos, 75, Botanikos
  GR-ATHENS (301)

- Agricultural Engineering Department,
  School of Agriculture, Aristotles University of Thessaloniki,
  GR-THESSALONIKI

- Biology Department,
  Lab. of General Botany,
  University of Athens,
  GR-ATHENS

- Institute of Viticulture, Vegetable Crops and Floriculture
  Department of Plant Residue and Agricultural Industries
  Residue Management and Treatment
  GR-HERAKLION (Crete)

- Laboratory of Organic Chemistry and Technology of the Technical
  University of Athens
  GR-ATHENS

6° IRELAND

Constructors & consultants :

- Biomass Ltd.,
  IRL-GALWAY

Laboratories & organizations :

- An Foras Taluntais (Kiely)
  Johnstown Castle
  IRL-Co-Wexford

- Microbiology Department (Colleran)
  University College of Galway
  IRL-GALWAY

- University College (Casey)
  IRL-DUBLIN

7° ITALY

Constructors & consultants :

a) Agricultural biogas plants

- ACTA s.r.l.
  Via Ajaccio, 12
  I-00198 ROMA

- Agipgiza s.p.a.
  Via Schwarin, 4
  I-42100 REGGIO EMILIA

- Biosystem srl
  Via Luciano Manara, 7
  I-20100 MILANO

- BSF
  Via M. Bartoli, 137
  I-00143 ROMA

- Clover
  C.P. 352
  I-47100 FORLI

- EBiA
  Via Bergamo, 21
  I-20100 MILANO

- Ecoplants
  Via Plauto, 3
  I-43026 S. LAZZARO (PR)

- EDAECO
  Via Verdi, 14
  I-26100 CREMONA

- Emmepi
  Via A. Volta
  I-37020 ARBIZZANO DI VALPOLICELLA (VR)

- Eurocom
  Via della Tecnica, 23
  S. Lazzaro di Savena
  I-40068 BO

- Eurodepuratori
  San Giacomo delle Segnate
  I-46020 MANTOVA

- FM-oraenergia
  I-42010 CACCIOLA, SCANDIANO (RE)

- Jacorossi
  Divisione Impianti Biogas
  Via Ostiense, 333/a
  I-00100 ROMA

- Rodon
  Via Flaminia, 161
  I-47037 RIMINI

- RPA
  Risorse Ambientali
  Strada del colle
  I-06074 FONTANA PERUGIA

- SES
  Via Cuneo, 20
  I-10152 TORINO

- SIDI
  Viale Mentana, 92
  I-43100 PARMA

- Snamprogetti
  Casella postale, 97
  I-61032 FANO (PS)

- Tecnoimpianti
  Via Maroni
  I-47020 MONTIANO (FO)

- Termomeccanica Italiana
  Via del molo, 1
  I-19100 LA SPEZIA

b) Industrial biogas plants

- Assoreni
  C.P. 15, Monterotondo
  I-00015 ROMA

- RPA
  Risorse Ambientali
  Strada del Colle
  I-06074 FONTANA PERUGIA

Laboratories & associations

- APRE
  Via Unione, 3
  I-20122 MILANO

- CRPA
  Via Crispi, 3
  I-42100 REGGIO EMILIA

- CSARE
  Corso del Popolo, 261
  I-ROVIGO

- CTIP Solar
  Piazzale G. Douhet, 31
  I-00144 ROMA

- EBiA
  Centro di Ricerche di Broni
  I-BRONI (PV)

- ENEA-FARE
  Centro Ricerche CASACCIA
  Strade Statale Anguillarese km 1.3
  I-00060 S. MARIA DI GALERIA, ROMA

- IRSA-CNR
  Reporto Sperimentale di Chimica e Tecnologia delle Acque
  Via F. De Blasio, 5
  I-70123 BARI

- Istituto di Chimica Industriale
  Universita di Padova
  Via Mazzolo, 9
  I-35100 PADOVA

- Istituto di Ingeneria Agraria
  Universita di Milano
  Via Celoria, 2
  I-20133 MILANO

- Istituto di Meccanica Agraria e Meccanizzazione
  Universita di Firenze
  P.le delle Cascine, 15
  I-50144 FIRENZE

- Istituto di Meccanica Agraria
  Universita di Padova
  Via Gravenigo, 6
  I-35100 PADOVA

- Istituto di Microbiologia Agraria
  Universita di Bologna
  Via Filippo Re, 6
  I-BOLOGNA

- Istituto di Microbiologia Agraria & Tecn.
  Universita di Firenze
  Piazzale Cascine, 27
  I-50144 FIRENZE

- Istituto di Principi di Ingegneria Chimica
  Universita di Napoli
  P.le Tecchio
  I-80125 NAPOLI

- Istituto di Scienze e Tecnologie dell'Ingegneria Chimica
  Universita di Genova

- Distillerie Orbat
  Via Roma, 15
  I-47034 FORLIMPOPOLI (FO)

8° THE NETHERLANDS

Constructors and consultants

- Fa. de Boer,
  Westerplantage, 15
  NL-8911 DC LEEUWARDEN

- Fa. van der Meijden,
  Retselseweg
  NL-   HEESWIJK-DINTHER

- Fa. van Wanrooij BV,
  Retselseweg, 6
  NL-   HEESWIJK-DINTHER

- Oostwouder BV
  Rijksweg, 126
  NL-   'T ZAND (N.H.)

- Paques BV
  T. de Boerstraat, 11
  NL-8561   EL BALK

b) Industrial biogas plants (UASB system)

- Centrale Suiker Maatschappij BV (Pette)
  Biothane Division
  Herengracht, 493
  NL-1017   BT AMSTERDAM

- Esmil Hubert BV
  Franekervaart
  NL-   SNEEK

- Grontmij NV (Lukkien/Postma)
  Consultants for Civil Engineering and Land Development
  Utrechtseweg, 62
  P.O. Box 153
  NL-3700   AD ZEIST

- Heidemij Adviesbureau (Zeevalkink)
  Lovinklaan, 1
  NL-6821   MX ARNHEM

- Technish Adviesbureau van de Unie van Waterschappen BV (Woelders)
  Teugseweg, 58026
  NL-7418   AM DEVENTER

Laboratories & associations :

- Agricultural University (Lettinga)
  Dept. of Water Pollution Control
  De Dreijen 12
  NL-6700   WAGENINGEN

- IMAG
  Institute for Mechanization, Labour and Construction (Poelma)
  P.O. Box 43
  NL-6700   WAGENINGEN

- Institute for Storage and Processing
  of Agricultural Produce (IBVL) (Rijkens)
  Bornsesteeg, 59
  P.O. Box 18
  NL-6700 WAGENINGEN

- MT-TNO (Colon)
  Applied Environmental Research Institute
  NL-APELDOORN

- University of Amsterdam

- University of Delft

- University of Groningen

- University of Nijmegen (D. Vogels)
  Faculty of Science
  Department of Microbiology
  Toernooiveld
  NL-6525 ED NIJMEGEN

9° UNITED KINDOM

Constructors & consultants :

a) Agricultural biogas plants

- Energy and Waste Systems Ltd. (BOVIS)
  Bridge House,
  UK-WESTBURY, Wiltshire BA134HS

- Farm Gas Ltd.
  Industrial Estate
  Love Lane
  UK-BISHOPS CASTLE, Shropshire

- Farrow/Wright Rain Ltd.
  UK-RINGWOOD, Hampshire

- Hamworthy Engineering Ltd.
  Pump and Compressor Division
  Fleets Corner,
  UK-POOLE, Dorset BH177LA

- A.O. Smith Harvestore Ltd.
  UK-EYE, Suffolk

- Helix Multiprofessional Services
  Mortimer Hill
  UK-MORTIMER, Reading

- Kilbees Slurry Digesters Ltd.
  Winkfield
  UK- WINDSOR FOREST, Berks

- Lockstoke Developments Ltd.
  Eastbourne Lane
  UK-MIDHURST, Sussex

- SCS Biotechnology Ltd.,
  Elmsfield
  UK-CHIPPING NORTON, Oxon

b) Industrial biogas plants

- Ames Crosta Babcock Ltd.
  UK-HEYWOOD, Lancashire OL10 2DX

- Anaerobic Digestion and Biomass Advancement Company Ltd. (ADBAC)
  (consultants)
  The Great Yard
  Marridge Hill, Ramsbury
  UK-MARLBOROUGH, Wilts

- Bio-energy International Ltd.
  Sandy Lane Industrial Estate
  UK-STOURPORT-ON-SEVERN, Worcestershire DY139QF

- Biomass International Ltd.
  P.O. Box 5
  UK-MILNTHORPE, Cumbria

- Biomechanics Ltd.
  Caxton House, Wellsley Road
  UK-ASHFORD, Kent TN7 8ND

- Cardiff Laboratories for Energy & Resources Ltd. (CLEAR) (consultants)
  UK-CARDIFF

- Cooper, MacDonald & partners (consultants)
  Loxford House, East Street
  UK-EPSOM, Surrey KT17 111G

- Dunlop Bioprocesses Ltd.
  Kingsbury Road
  UK-ERDINGTON, Birmingham B24 9QU

- Jones and Attwood Ltd.
  Titan Works
  UK-STOURBRIDGE, West Midlands, DY8 4LR

- Water Research Centre (consultants)
  Stevenage Laboratory
  Elder Way
  UK-STEVENAGE, Herts SG1 1TH

c) Landfills

- Dept. of Public Health Engineering
  Greater London Council, County Hall
  UK-LONDON SW1

- Environmental Safety Group
  AERE
  UK-HARWELL, Oxfordshire OX110RA

- London Brick Landfill Ltd.
  UK-STEWARTBY, Bedford MK43 GL2

d) Gas use

- Bristol Pneumatic Ltd.,
  Lodge Causeway
  UK-FISHPONDS, Bristol BS16 3JS

Laboratories & organizations

- BABA (British Anaerobic and Biomass Association)
  The White House, Little Bedwyn
  UK-MARLBOROUGH, Wilts SN8 3JP

- British Agricultural Export Council
  Belgrave Square, 35
  UK-LONDON SW1

- British Water and Effluent Treatment Plant Association (BWEPTA)
  Castle St., High Wycombe, 51
  UK-BUCKS

- Food Research Institute
  Comery,
  UK-NORWICH, Norfolk

- Leicester Polytechnic School of Mechanical & Production Engineering
  (Picken)
  P.O. Box 143
  UK-LEICESTER LE1 9BH

- National Institue for Research in Dairying (NIRD)
  Shinfield
  UK-READING, Berks

- Pollution Research Unit (Wheatley)
  UMIST
  P.O. Box 88
  UK-MANCHESTER M6010D

- Rowett Research Institute (Hobson)
  Microbiology Department,
  Greenburn Road,
  Bucksburn
  UK-ABERDEEN AB2 9SB

- The Polytechnic of Wales (Hawkes)
  Department of Science
  Anaerobic Digester Research & Development Unit
  UK-PONTYPRIDD, Mid Glamorgan CF 371DL (Wales)

- University College
  Department of Microbiology (Stafford)
  Newport Road
  P.O. Box 78
  UK-CARDIFF CFI 1XL (Wales)

- Warren Spring Laboratory (WSL) (Le Roux)
  Gunnels Wood Road
  Stevenage
  UK-HERTS SG1 2BX

10° SWITZERLAND

Constructors & consultants :

- Arnold AG
  Gülle- und Farmtechnik
  CH-6203 SEMPACH-STATION

- Binotech AG
  CH-8325 ROGGWIL

- Buri & Cie. SA
  Place Cornavin, 12
  CH-1201 GENEVE

- Cetec AG Turgi
  Schiffmühle
  CH-5300 TURGI

- J.-M. Chapallaz
  CH-1445 VUITEBOEUF

- Eberle & Frick AG
  Bodahus, 493
  CH-EL-9497 TRIESENBERG

- Energies
  Bureau d'ingénieurs
  J. Olsommer
  Case postale 100
  CH-1870 MONTHEY

- CJ Energies Nouvelles SA
  Route de Genève, 82
  CH-1000 LAUSANNE 20

- R. Epars
  Chemin des Clochettes, 12
  CH-1206 GENEVE

- EREP SA
  La Rosaire
  CH-1111 ACLENS

- W. Huber AG
  Kunststoffwerk
  CH-5426 LENGNAU

- Inventa AG
  CH-7013 DOMAT/EMS

- M. Lack
  Rocher 15
  CH-2000 NEUCHATEL

- LBA
  Laurstr., 10
  CH-5200 BRUGG

- R. Manahl
  Bergstr., 8
  CH-A-6900 BREGENZ

- Pfulg F.
  Ch. Nant-d'Argent, 20A
  CH-1223 COLOGNY

- J. Python
  Route Henry Dunant, 1
  CH-1700 FRIBOURG

- Rotaver AG
  Kunststoffwerk
  CH-3432 LÜTZELFLÜH

- W. Rothen
  Tisserands, 34
  CH-1510 MOUDON

- Schmid AG
  System Reusch
  CH-8360 ESCHLIKON

- R. Schwab
  Murtenstrasse
  CH-3210 KERSERS

- SEDE SA
  Rue du Midi, 33
  CH-1800 VEVEY

- M. Sieber
  Bungertweg
  CH-8450 ANDELFINGEN

- D. Wahl
  Witellikerstr., 43
  CH-8702 ZOLLIKON

Laboratories & associations

- EAWAG
  Eidgenössisches Anstalt für Wasserversargung und Abwasserreinigung und Gewüsserschutz (Schertenleib)
  Uberlandstrasse, 133
  CH-8600 DÜBENDORF

- Ecole Polytechnique Fédérale de Lausanne (EPFL) (Peringer)
  Chaire de Génie Biologique
  CH-1015 LAUSANNE

- Laboratorium für Milchwissenschaft (IIZH) (Favre)
  Institut für Lebenmittelwissenschaft
  ETH-Zentrum, Eisgasser, 8
  CH-8004 ZÜRICH

- SSES, Schweizerische Vereinigung für Sonnen Energie; Société Suisse pour l'énergie solaire
  Multschellenstrasse, 4
  CH-8002 ZÜRICH

- SVLT, Schweizerische Verband für Landtecknik, or :
  ASETA, Association Suisse pour l'Equipement Technique dans l'Agriculture
  Postfach
  CH-5223 RINIKEN/AG

- Swiss Federal Institute of Technology ETH (Woschitz)
  (Eidgenössische Technische Hochschule)
  Projekt Biogas, ETH-Zentrum
  Clausiusstrasse, 47
  CH-8092 ZÜRICH

- Swiss Federal Research Station for Agricultural Chemistry and Environmental Hygiene, FAC (Eidgenössische Forschungsanstalt für Agrikulturchemie und Umwelthygiene) (Besson)
  CH-3097 LIEBEFELD

- Swiss Federal Research Station for Farm Management and Agricultural Engineering, FAT (Eidgenössische Forschungsanstalt für Betriebswirtschaft und Lantechnik), (Kaufmann, Wellinger)
  CH-8355 TÄNIKON

- Université de Neuenburg
  Laboratoire de Microbiologie
  Institut de Botanique
  Rue Chantemorte, 22
  CH-2000 NEUCHATEL

APPENDIX V. SYMBOLS AND UNITS

Symbols and units used throughout the survey are based on two proposals. The first is entitled "Proposal for the Definition of Parameters and Analytical Measurements Applicable to the Anaerobic Digestion Process". It has been proposed to the Commission of the European Communities by a team of experts (F. Colin, G.H. Ferrero, M. Gerletti, P. Hobson, P. L'Hermite, H.P. Naveau and E.-J. Nyns). This document, reference DOC XII MPS 15/82-EN 10.2.83 Rev 3, has been published in Agricultural Wastes (Colin et al., 1983). The second is entitled "Nomenclature and Symbols in Biological Wastewater Treatment" and was prepared for the International Association on Water Pollution Research (IAWPR) and the Commission on Water Quality of the International Union of Pure and Applied Chemistry (IUPAC) by a team of experts (P. Grau, P.M. Sutton, M. Henze, S. Elmaleh, C.P. Grady, W. Gujer and J. Koller). This document is in trial use for one year. It will be published in Water Research.

Table 5.1 lists the symbols and their names and gives dimensions and units. Table 5.2 lists complementary abbreviations. Where necessary, more subscripts, including abbreviations are used. E.g. B_{VCOD} is the volumetric loading rate in kg COD per unit digester working volume per day; $Y_{S\ VSr/S\ VSo}$ is the conversion expressed in volatile solids.

Table 5.1. Symbols and units

Symbol	Name	Dimension	Units	
			MKS	Conventional
S	Concentration in substrate biomass (feedstock)	ML^{-3}	$kg\ m^{-3}$	$kg\ m^{-3}$
X	Concentration in active biomass (microorganisms)	ML^{-3}	$kg\ m^{-3}$	Various
Subscripts : o e r	Influent Effluent Removed			
B_V	Volumetric loading rate	$ML^{-3}T^{-1}$	$kg\ m^{-3}\ s^{-1}$	$kg\ m^{-3}\ d^{-1}$
B_X	Food to microorganism ratio	$ML^{-3}T^{-1}$	$kg\ m^{-3}\ s^{-1}$	(various) $m^{-3}\ d^{-1}$
r_V (%)	Volumetric biogas production rate (specify methane content)	T^{-1}	$m^3\ m^{-3}\ s^{-1}$	$m^3\ m^{-3}\ d^{-1}$
r_X	Specific methane production rate	L^3T^{-1} (various)$^{-1}$	m^3 (various)$^{-1}$ s^{-1}	m^3 (various)$^{-1}d^{-1}$
$Y_{CH4/So}$ gas/	Yield : methane produced per unit biomass substrate introduced in the methane digester	$L^3\ M^{-1}$	$m^3\ kg^{-1}$	$m^3\ kg^{-1}$
$Y_{Sr/So}$	Conversion : biomass substrate removed per biomass substrate introduced in the methane digester	-	$kg\ kg^{-1}$	$kg\ kg^{-1}$
θ	Mean (hydraulic) detention time	T	s	d

Table 5.2. Abbreviations

Abbreviation	Name	Units
COD	Chemical oxygen demand	mg O_2 l^{-1}
TS	Total solids (dry weight : 105 °C)	
VS	Volatile solids (weight at 105 °C less weight either at 450 °C or 600 °C)	
Subscripts : H	Hydraulic	
ML	Digester working volume (Mixed liquor)	m^3
S	Solids	
ECU	European currency unit	1 ECU = ± 0.9 US $
o.e.	oil equivalent	Mt (10^6 tons)

GLOSSARY

ACETOCLASTIC : anaerobic bacteria which dismutate acetate into methane and carbon dioxide.

ACTIVE BIOMASS : biomass or part thereof which has a (micro)-biological activity.

ALKALINITY : in the context of the present book, this parameter identifies the "buffering" capacity of a mixed liquor. It takes into account the volatile fatty acid anions and bicarbonate, HCO_3^- and their corresponding cations, namely ammonium, NH_4^+.

ANAEROBIC CONTACT SYSTEM : continuous, completely mixed microbiological process with recycle of the active biomass decanted in a separate sedimentation tank.

ANAEROBIC DIGESTION : see Biomethanation.

ANAEROBIC FILTER SYSTEM : up- or downflow continuous microbiological process with trapping and immobilization of the active biomass on a fixed carrier.

BACTERIAL COMMUNITY : group of bacteria including several bacterial populations.

BACTERIAL POPULATION : group of bacteria belonging to the same species, exchanging genetic material among each other and having few contacts with other groups of the same species.

BATCH SYSTEM : discontinuous microbiological system characterized by a single loading at the beginning of the process and a single unloading at the end of the process. In the variant fed batch system, the single loading is replaced by multiple step loading.

BIOCHEMICAL OXYGEN DEMAND (BOD) : amount of oxygen (as mg l^{-1}) required by bacteria to degrade that part of the organic matter, particularly in (waste)water which is biodegradable in a fixed period of time (usually 5 days).

BIOCONVERSION : transformation of matter by living organisms.

BIOFUELS : energy fuels obtained by bioconversion of biomass.

BIOGAS : gas obtained by bioconversion of organic matter. Usually restricted to gases whose major component is methane. Biomethanation yields a biogas containing from 50 to 75 % methane, CH_4, from 35 to 40 % carbon dioxide, CO_2, and 0-5 % of various other gases such as sulphydric acid, H_2S, ammonia, NH_3, dihydrogen, H_2, and water vapour.

BIOGAS PLANT : a methane digester together with its ancillary systems, namely the storage and/or treatment system of the influent, the storage, and/or treatment and utilization system of the biogas, and the storage and/or treatment of the effluent.

BIOMASS : (1) organic matter of plant, microbial or animal origin, the transformation products and wastes or residues thereof; (2) sometimes restricted to wastes and residues of agricultural origin.

BIOMASS SUBSTRATE : see Feedstock.

BIOMETHANATION : anaerobic microbial process by which organic matter is bioconverted into biogas.

BIOMETHANATION SYSTEM : biogas plant integrated in an enterprise such as a farm, an industrial plant or a landfill.

CARRIER : matrix placed inside a reactor on which the active biomass is immobilized.

CHEMICAL OXYGEN DEMAND (COD) : amount of oxygen, as mg l^{-1}, used to oxidize organic matter chemically. It includes the BOD but is more than that.

C/N RATIO (carbon to nitrogen ratio) : ratio of carbon content to nitrogen content. In biomethanation, it characterizes a feedstock and relates to the reliability of the process.

COMPLETELY MIXED SYSTEM : abbreviated name for a continuous single-step, once-through microbiological process without active biomass recycle. Continuous or intermittent mixing homogenizes the mixed liquor. Loading and equivalent unloading of the reactor can occur continuously or intermittently.

DETENTION TIME : see Mean detention time.

DIGESTER : see Methane digester.

ECOSYSTEM : biological system including its physical and chemical environmental factors.

EFFLUENT : (1) mixed liquor flowing out of a reactor tank; (2) liquid with or without suspended particles or solids discharged from a biogas plant (in the context of the present book).

ENERGY CROP : selected plants grown and harvested as biomass to be used as an energy source or converted into an energy fuel or product.

EXO-ENZYME : enzyme located on the outer side of the bacterial cell membrane, in the space between the cell wall and the cell membrane, or released by the bacteria in the outer medium.

EXPANDED BED SYSTEM : see Fluidized bed system.

FACULTATIVE ANAEROBIC : bacteria which can live either in the presence or in the absence of oxygen.

FED BATCH SYSTEM : see Batch system.

FEEDSTOCK (syn. biomass substrate) : matter used to load a reactor. See also Biomass.

FERMENTATION : (1) anaerobic biological redox transformation of organic matter which does not involve oxygen-containing mineral compounds; (2) technological process based on (1). Often characterized by the end product.

FERMENTATIVE : micro-organism competent to perform fermentation.

FIXED BED SYSTEM : see Anaerobic filter system.

FIXED FILM : thin layer of active biomass immobilized on a fixed carrier.

FLOATING GASHOLDER : see Gasholder.

FLOC : (loose) association of suspended solids composed of particles ranging in size up to a few mm; this term is especially applied to bacteria.

FLOCCULATION : physico-chemical process by which suspended solids assemble in a floc.

FLUIDIZED (EXPANDED) BED SYSTEM : upflow (microbiological) process in which the active biomass is immobilized on micro-carriers which in turn are expanded by liquid recirculation at low superficial velocities or are fluidized by liquid recirculation at high superficial velocities.

GASHOLDER : tank for the storage of (bio)gas. A floating gasholder is a hollow cylinder closed at its upper end and sealed at its lower end by a liquid, generally water, contained in a tank; the gas is stored at low pressure within the cylinder above water level, the cylinder rising or falling to accommodate the varying gas volume.

GAS LIFT : mixing device consisting of a vertical draft tube immersed in a liquid, from the lower end of which compressed gas is expelled.

HYDROGENOTROPHIC : anaerobic bacteria able to reduce carbon dioxide with dihydrogen to methane.

HYDROLYSIS : catalytic breaking down of large molecules into smaller ones with water as reagent. In the context of this book, the catalysts are often bacterial exo-enzymes.

IMMOBILIZATION : physico-chemical process by which bacteria adhere, or are fixed to a carrier.

INFLUENT : (liquid) feedstock loaded into a reactor tank.

INTERNAL RATE OF RETURN : discount rate that equates the present value of the expected net future cash-flows or receipts to the initial cash outlay.

IRIS SYSTEM : similar to the anaerobic contact system, but in one tank.

LANDFILL : site where domestic refuse is discharged.

LIQUEFACTION : physico-chemical and biological process by which solid (organic) matter changes into soluble matter.

LIQUID or SEMI-SOLID MANURE (syn. manure) : animal waste from intensive farms where no bedding is used; usually collected as a slurry.

LOADING : process by which a feedstock is introduced into a reactor tank. Loading can take place continuously, intermittently (i.e., for given lengths of time or at given intervals of time), or discontinuously; in the latter case, loading can be once only or step-wise.

MANURE : see Liquid or semi-solid manure.

MANURE WITH BEDDING : refuse of stables and barnyards consisting of livestock excreta with litter.

MEAN DETENTION TIME : mean amount of time that each unit volume of mixed liquor is retained in a tank. The hydraulic mean detention time refers to the liquid portion of the mixed liquor. The solids mean detention time refers to the solid portion of the mixed liquor.

MESOPHILIC : descriptive of biological processes taking place at temperatures ranging from 30° to 40 °C.

METHANE DIGESTER (syn. digester) : reactor or closed vessel in which organic wastes are bioconverted into methane and carbon dioxide under anaerobic conditions.

METHANOGENESIS : final step of the biomethanation process during which acetate, dihydrogen and bicarbonate are bioconverted into methane and carbon dioxide.

MIXED LIQUOR : mixture of feedstock and active biomass in the reactor tank or its effluent.

NET PRESENT VALUE : actualized value of net future cash-flows.

PARTIALLY MIXED SYSTEM : differs from the completely mixed system by non-complete homogenization of the digester content.

PAY-BACK PERIOD (SIMPLE) : number of years necessary for a project to recover its original investment from the expected net cash-flows.

PERCOLATION : passage of a liquid through particulate solid matter. Used as pretreatment or first step of the biomethanation process.

PLUG-FLOW SYSTEM : single step, once-through system with horizontal flow of the mixed liquor which can be mechanically assisted.

PRETREATMENT : processing of a feedstock prior to its treatment.

PSYCHROPHILIC : descriptive of biological processes taking place at ambient temperatures and below.

RESIDUE : left-over of an activity which has a positive value. Straw is an agricultural residue.

SCUM : layer of solids at the surface of the mixed liquor, notably when mixing is not sufficient.

SETTLER : tank in which a mixture of solids and liquids such as a feed-stock or an effluent is treated by sedimentation to clarify the liquid and thicken the settleable solids.

SETTLER GAS SEPARATOR : settler designed to separate a gas, namely biogas, from a liquid, namely mixed liquor.

SLUDGE BED : layer of active biomass in a tank reactor. See Upflow system without carrier.

SOURING : occurs when the biological activity (in a methane digester) switches from neutral methanogenesis to acid fermentation.

START-UP : first period of operation of a digester during which it is loaded, seeded with active biomass and progressively brought to steady running conditions.

THERMOPHILIC : descriptive of biological processes taking place at temperatures of 50 °C upwards.

TWO-STEP SYSTEM : liquefaction and biomethanation in two separate reactors in series. Acidogenic fermentation occurs in the first reactor, whereas methanogenesis occurs in the second one, each reactor being designed accordingly.

UPFLOW SYSTEM WITHOUT CARRIER : upflow system with an incorporated settler gas separator. The active biomass remains trapped in the reactor as a sludge bed.

VOLATILE FATTY ACIDS : aliphatic acid with 2 to 5 carbon atoms. A classical end product in the decomposition of organic matter by fermentative bacteria.

VOLATILE SOLIDS : organic matter estimated as loss by ignition of the dry solids at temperatures ranging from 450° to 600 °C for periods of time ranging from 3 to 12 h.

WASTE : left-over of an activity which has a negative value. Wastewater is a typical waste.

REFERENCES

Anon. a., Impianti di digestione anaerobica. Istruzioni per la construzione e la conduzione. Italian Norm UNI 005.

Anon. b., A code of practice on safety in and around anaerobic digesters. British Anaerobic and Biomass Association, Ltd. (BABA), ed. The White House, Little Bedwyn, Marlborough, Wilts., United Kingdom.

Anon. (1978), FAST Programme of the Commission of the European Communities. Official Journal, L225, 38 (16.08.78)

Anon. (1979), Construction et exploitations d'installations de biogaz. Complément provisoire 1979 à la partie des "Directives Installations de Chauffage". Association des Etablissements Cantonaux d'Assurance contre l'Incendie, Suisse.

Anon. (1981a), Report of the Synthesis Group. UN Conference on New and Renewable Sources of Energy. Nairobi 1981, Doc. A/CONF. 100/PC/41. March 10.

Anon. (1981b), Report of the Biomass Panel. UN Conference on New and Renewable Sources of Energy. Nairobi 1981, Doc. A/CONF. 100/PC/28.

Anon. (1982a). "Biotechnology in Europe". European Federation of Biotechnology, ed. DECHEMA, publ., Postfach 970146, D-6000 Frankfurt (Main) 97, Germany.

BAADER, W., and KLOSS, R. (1982). Assessment of biogas installations in the Federal Republic of Germany. Report to the Commission of the European Communities. Project Reference n° ESE-R-051-D (N).

BEHRENS, D., and BUCHHOLZ, K., (1982) eds. A community strategy for European biotechnology. DECHEMA, publ., Postfach 970146, D-6000 Frankfurt (Main) 97, Germany.

BINOT, R.A., BOL, T., NAVEAU, H.P. and NYNS, E.-J. (1982). Biomethanation by immobilised fluidised cells. Water Science Technology, 15, Copenhagen, 103-115.

CANTLEY, M. (1980) ed. FAST, Sub-Programme "Bio-Society". Doc. EUR 7105. Commission of the European Communities, publ., DG XII, Brussels.

CHARTIER, P. (1983). Summing up and evaluation of biomass prospects, In Energy from Biomass. STRUB, A., CHARTIER, P., and SCHLESER, G., eds. Applied Science publ., London, United Kingdom, VII-IX.

COLIN, F., FERRERO, G.L., GERLETTI, M., HOBSON, P., LHERMITE, P., NAVEAU, H.P., and NYNS, E.-J. (1983). Proposal for the definition of parameters and analytical measurements applicable to the anaerobic digestion processes. Agricultural Wastes, 7, 183-193.

DEVILLERS, P., LESCURE, J.P., and BOURLET, P. (1977). Nouvelles perspectives pour le traitement des eaux en sucrerie. La fermentation mésophile. Sucrerie Française (Avril) 173-183.

DUCELLIER, G., and SAUZE, F. (1975). Le méthane biologique. La France Viticole, November 1975 to March 1976.

DUCELLIER, G., and ISMAN, M. (1941). Premier aperçu sur le gaz du fumier. L'Agria, 38 (Mars).

HALL, D.O. (1983). Food versus fuel. A world problem, In Energy from Biomass. STRUB, A., CHARTIER, P. and SCHLESER, G., eds. Applied Science publ., London, United Kingdom, pp 43-62.

HASHIMOTO, A.G. (1983). Proceedings IIId International Symposium Anaerobic Digestion, Boston, Mass.

LAGRANGE, B. (1979). Biométhane, Tome 1 : "Une Alternative Crédible". Tome 2 : "Principes, Techniques, Utilisations". Edisud, Aix-en-Provence, France.

LEE, C., and McKIM, T. (1981). Water hyacinth wastewater treatment system, In Energy from Biomass and Wastes. Institute of Gas Technology, ed. Chicago, Ill., U.S.A., 59-86.

LEGROS, A., ASINARI di SAN MARZANO, C.-M., NAVEAU, H.P. and NYNS, E.-J. (1983). Fermentation profiles in bioconversions. Biotechnology Letters, 5 (1), 7-12.

LETTINGA, G., VAN VELSEN, A.F.M., HOBMA, S.W., DE ZEEUW, W., and KLAPWIJK, A. (1980). The use of the up-flow sludge blanket (USB) reactor concept for biological wastewater treatment, especially for anaerobic treatment. Biotechnology and Bioengineering, 22, 699-734.

McCARTY, P.L. (1982). One hundred years of anaerobic treatment. In Anaerobic Digestion 1981. HUGHES, D.E., STAFFORD, D.A., WHEATLEY, B.I., BAADER, W., LETTINGA, G., NYNS, E.-J., VERSTRAETE, W. and WENTWORTH, R.L., eds. Elsevier Biomedical Press, Amsterdam.

NYNS, E.-J., NAVEAU, H.P., CHOME, R. and BERTRAND, Y. (1979). Digesters. A worldwide review. In Anaerobic Digestion. STAFFORD, D.A., WHEATLEY, B.I., and HUGHES, D.E., eds. Applied Science publ. London, United Kingdom, 37-59.

NYNS, E.-J. (1981). Etat de la biochimie et de la microbiologie de la digestion méthanique. Tribune du Cebedeau, 34, 351-356.

NYNS, E.-J., DEMUYNCK, M., NAVEAU, H.P., RAO, K.K., and HALL D.O. (1982). Energy (biomethanation and photobiology). In A Community Strategy for European Biotechnology, FAST - Bio-Society. BUCHHOLZ, K., European Federation of Biotechnology, eds. DECHEMA, publ., Postfach 970146, D-6000 Frankfurt (Main) 97, Germany.

PALZ, W., and CHARTIER, P. (1980) eds. Energy from Biomass in Europe. Applied Science publ., 234 pp.

RIJKENS, B.A. (1981). Two-phase process for the anaerobic digestion of organic wastes yielding methane and compost, in Energy from Biomass, vol. 1. Chartier, P., and Palz, W., eds. D. Reidel Publishing Company, Dordrecht, The Netherlands.

SALKINOJA-SALONEN, M., HAKULINEN, R., VALO, R., and APAJALAHTI, J. (1982). Biodegradation of recalcitrant organochlorine compounds in fixed film reactors. Preprints of papers presented at the IAWPR specialized seminar "Anaerobic Treatment of Wastewater in Fixed Film Reactors", June, Copenhagen, Denmark.

SAUZE, F. (1947). Le gaz de fumier. L'Agriculture pratique, octobre.

STAFFORD, D.A., HAWKES, D.L., and HORTON, R. (1980). Methane Production From Waste Organic Matter. C.R.C. Press, Boca Raton, Florida, U.S.A.

STEFFEN, A.J., and BEDKER, M. (1962). Operation of full scale anaerobic contact treatment plant for meat packing wastes. Proceedings, 16th Industrial Waste conference, Purdue University, West Lafayette, Ind. U.S.A., 423-437.

SWITZENBAUM, M.S., and JEWELL, W.J. (1980). Anaerobic attached film expanded bed reactor treatment. Journal of the Water Pollution Control Federation, 52 (7), 1953-1965.

TABASARAN, O. (1982). Obtention et valorisation du methane à partir de déchets urbains. La Tribune du Cebedeau, 35 (469) 483-488.

VAN BRAKEL, J. (1980). The Ignis Fatuus of Biogas. Delft University Press, Delft, The Netherlands.

WISE, D.L., ed. (1981). Fuel Gas Production From Biomass. C.R.C. Press, Boca Raton, Florida, U.S.A., Vol. I. and II.

YOUNG, J.C. and McCARTHY, P.L. (1969). The anaerobic filter for waste treatment. Journal of the Water Pollution Control Federation, 41 (5), R160-R173.

WELLINGER, A. (1982). Erfahrungen mit Biogasanlagen. Proceedings Fachtagung Biogas 1982, A. Wellinger, Edelmann W., Haltiner E., eds. SSIV and NEFF. pp. 34-55.

SUBJECT INDEX

Printed in the United States
88362LV00003B/43/A

9 789027 717801